THE WINES OF SOUTHWEST U.S.A.

THE CLASSIC WINE LIBRARY

Editorial board: Sarah Jane Evans MW,
Richard Mayson and James Tidwell MS

There is something uniquely satisfying about a good wine book, preferably read with a glass of the said wine in hand. The Classic Wine Library is a series of wine books written by authors who are both knowledgeable and passionate about their subject. Each title in The Classic Wine Library covers a wine region, country or type and together the books are designed to form a comprehensive guide to the world of wine as well as an enjoyable read, appealing to wine professionals, wine lovers, tourists, armchair travellers and wine trade students alike.

Port and the Douro, Richard Mayson
Cognac: The story of the world's greatest brandy, Nicholas Faith
Sherry, Julian Jeffs
Madeira: The islands and their wines, Richard Mayson
The wines of Austria, Stephen Brook
Biodynamic wine, Monty Waldin
The story of champagne, Nicholas Faith
The wines of Faugères, Rosemary George MW
Côte d'Or: The wines and winemakers of the heart of Burgundy,
    Raymond Blake
The wines of Canada, Rod Phillips
Rosé: Understanding the pink wine revolution, Elizabeth Gabay MW
Amarone and the fine wines of Verona, Michael Garner
The wines of Greece, Konstantinos Lazarakis MW
Wines of the Languedoc, Rosemary George MW
The wines of northern Spain, Sarah Jane Evans MW
The wines of New Zealand, Rebecca Gibb MW
The wines of Bulgaria, Romania and Moldova, Caroline Gilby MW
Sake and the wines of Japan, Anthony Rose
The wines of Great Britain, Stephen Skelton MW
The wines of Chablis and the Grand Auxerrois, Rosemary George MW
The wines of Germany, Anne Krebiehl MW
The wines of Georgia, Lisa Granik MW
The wines of South Africa, Jim Clarke
The wines of Southwest U.S.A., Jessica Dupuy

THE WINES OF
SOUTHWEST U.S.A.

A guide to New Mexico, Texas, Arizona, and Colorado

JESSICA DUPUY

ACADEMIE DU VIN LIBRARY

Jessica Dupuy is a freelance journalist who has covered wine and food for more than a decade. She has written about Texas wine, specifically for *Texas Monthly* magazine since 2007. In addition, she is a regular contributor as a wine and food columnist for Forbes.com and has written for *Imbibe*, GuildSomm, *SevenFifty Daily*, *Wine Enthusiast*, *Southern Living*, and *National Geographic Traveler*. She is a Certified Sommelier through the Court of Master Sommeliers, an Advanced Certificate holder with the Wine and Spirits Educational Trust, and a Certified Specialist of Wine and Spirits through the Society of Wine Educators.

Jessica has also covered food trends for various publications throughout her career and is the author of *Uchi: The Cookbook*, *The Salt Lick Cookbook: A Story of Land, Family and Love*, *Jack Allen's Kitchen Cookbook*, *The United Tastes of Texas* and *The United Tastes of the South* through *Southern Living* magazine, and *Tex-Mex: Traditions, Innovations, and Comfort Foods from Both Sides of the Border*.

Dupuy holds a B.A. in History from Trinity University and an M.A. in Journalism from the Medill School of Journalism at Northwestern University. She is a member of Les Dames D'Escoffier–Austin. Dupuy lives in the Hill Country town of Dripping Springs, just west of Austin with her family. Among the things she enjoys most are cooking with her kids, Gus and Ashlyn, sharing great wine with friends, and fly-fishing with her husband, Myers. Follow her on Instagram at @jessicandupuy and on Twitter at @JessicaNDupuy.

Copyright © Jessica Dupuy, 2020, 2024

The right of Jessica Dupuy to be identified as the author of this book has been asserted in accordance with the Copyright, Designs and Patents Act 1988.

First published in 2020 by Infinite Ideas Limited
This edition published 2024 by Académie du Vin Library Ltd
academieduvinlibrary.com

All rights reserved. Except for the quotation of small passages for the purposes of criticism or review, no part of this publication may be reproduced, stored in a retrieval system or transmitted in any form or by any means, electronic, mechanical, photocopying, recording, scanning or otherwise, except under the terms of the Copyright, Designs and Patents Act 1988 or under the terms of a licence issued by the Copyright Licensing Agency Ltd, 90 Tottenham Court Road, London W1T 4LP, UK, without the permission in writing of the publisher.

A CIP catalogue record for this book is available from the British Library

ISBN 978–1–913141–73–8

Brand and product names are trademarks or registered trademarks of their respective owners.

Front cover: © Kenny Braun
Back cover: Lescombes Family Vineyards

Photos: D.H. Lescombes plate 1 (top); La Viña Winery plate 1 (bottom); Noisy Water Winery plate 2 (top), p. 9; Vivác Winery plate 2 (bottom), pp. 28 and 40; Lost Draw Cellars plate 3; Jessica Dupuy plate 4 (top), p. 73; Pedernales Cellars plate 4 (bottom); An Pham plate 5 (top), pp. 129 and 148; Jennelle Bonifield plate 5 (bottom), p. 156; Callaghan Vineyards plate 6 (top); Newport Photography plate 6 (bottom), p. 151; Armando Martinez plate 7 (top), plate 8, p. 195; CraDurr Photography plate 7 (bottom); Kurt Weddle p. 53; Yavapai College p. 174.

Maps by T.J. Tucker

Printed in Great Britain

CONTENTS

Introduction	1
Part 1: New Mexico—where the southwest begins	9
1. History	11
2. Climate, regions, and challenges	21
3. New Mexico producers	31
4. Where to eat and stay in New Mexico	43
Part 2: Texas—wine in the Lone Star State	53
5. History	55
6. Climate, regions, and challenges	65
7. Texas producers	81
8. Where to eat and stay in Texas	121
Part 3: Arizona—desert wine	129
9. History	131
10. Climate, regions, and challenges	141
11. Arizona producers	155
12. Where to eat and stay in Arizona	183
Part 4: Colorado—high altitude wine	195
13. History	197

viii   THE WINES OF SOUTHWEST U.S.A.

14. Climate, regions, and challenges	203
15. Colorado producers	213
16. Where to eat and stay in Colorado	229
Bibliography	239
Acknowledgments	243
Index	245

INTRODUCTION

In conjuring up the American Southwest, the imagination swarms with visions of arid, windswept terrain punctuated by vast ranches framed by barbed-wire fences and scattered with grazing cattle. The region evokes images of terraced rock formations and jagged mountain peaks towering in an array of vibrant, earthen reds and browns against the backdrop of radiant blue skies, with sunsets that fade these radiant colors into a brilliance of pastel pinks, oranges, and purples. It's the birthplace of the American frontier, a playground for outdoor adventurers, and the inspiration for countless watercolor canvases and vivid oil-painted landscapes. Indeed, this romantic imagery has swayed myriad settlers to the region for centuries. And while few would have ever associated this enchanting region with wine, it was here that the first *Vitis vinifera* vines in the United States were planted. However, it wasn't until the 1970s that the modern wine industry of the American Southwest was born. By and large, Texas has led the pack in terms of overall growth in vineyard plantings and production since then. But New Mexico, Arizona, and Colorado have also evolved, with compelling stories and promising wine industries of their own.

I have to admit, taking on the project to share the story of wine in a quadrant of America that is most associated with cacti, cattle, and cowboys was a bit daunting. It's a task I met with conflicting emotions. On the one hand, I was ecstatic to have an opportunity to share with readers a corner of the country that I've called home for my entire life. As a fifth-generation Texan from the central Hill Country area of the state, I've traveled to all corners of its borders and continue to marvel at its vast geographical and cultural diversity. In my lifetime, I've clocked thousands of miles on

2 THE WINES OF SOUTHWEST U.S.A.

various vehicle odometers on repeated road trips throughout New Mexico, Arizona, and Colorado. (I even spent a couple of summers as a horse wrangler on a Colorado ranch, but that's a story for another book.)

Inspired by its endless recreational outlets—from hiking and fly-fishing to white-water rafting and skiing—the Southwest continues to lure me out for seasonal adventures with my own young family. The rich, one might say romantic, history of the wild west is what led me to focus my studies on history in college, and later capture the region's unique, deeply-layered flavors in a few different cookbooks. In short, the American Southwest holds a very special place in my heart.

However, the task also initially filled me with dread. Sharing the story of the wine regions of four very different states is a lofty endeavor. They each have their own distinct successes and challenges. But as is the case for other iconic wine regions around the world, you can't truly understand the wine narrative without having a grasp on the history and culture. Indeed, boiling down the past five centuries of four very different states is no easy task.

But that wasn't the biggest concern. The Southwest United States covers a massive area. Texas alone is larger than all of France by about 20,000 square miles (51,800 square kilometers). When you add the other three states, you're looking at an area that stretches for more than 1,300 miles (2,100 kilometers) from east to west as well as north to south. (It would take roughly 20 hours, without stopping, to drive east to west or north to south.) From north to south, it's about the same distance from the French beaches of Normandy to the bottom of the boot heel of Italy, or Porto, Portugal to Munich, Germany. To say it's a lot of ground to cover is an understatement.

Ultimately, I felt up to the task. The best part about covering a vast region that has been largely under-reported is introducing others to the topography of the lesser-known viticultural areas, the very elegant and finessed wines that these places produce, and the faces of the earnest, hard-working people who make them.

One poignant thing I discovered while researching this book is that I'm not the only one persuaded by the potential for great wine in the Southwest. As it turns out, many of the top winemakers in the region have sought wine education and winemaking experience in iconic wine regions not only throughout California but also in Burgundy and the Rhône Valley in France, Friuli and the Veneto in Italy, Spain, New

Zealand, Australia, and Chile. Experience in these regions brought discernment and wisdom to this stable of producers, yet ultimately a desire to pursue the great potential in their home region drew them back to the Southwest. Perhaps another common thread worth noting is the prevalence of native and hybrid grape varieties being grown throughout each state. In Texas, Blanc Du Bois and Black Spanish have proven their merit against Pierce's disease. In the colder climes of northern New Mexico and mountainous Colorado, cold-hardy varieties such as Chambourcin, Marquette and La Crescent have experienced a surge. Though *vinifera* continues to represent the majority of plantings, this trend in native and hybrid varieties reveals a prudent awareness of the need to focus on those varieties that work best within the region's unique growing conditions. While it's true that this book leaves a lot left unsaid, my hope is that it at least digs deep enough beneath the surface to pique curiosity to seek out the wines of the American Southwest.

Foundational dates of the American Southwest

1200s	Acoma Pueblo in New Mexico, the longest inhabited community in the U.S.
1300s	Mesa Verde and many other Anasazi sites abandoned in Southwest Colorado
1520	Cortez arrives in Mexico
1540	Coronado crosses into current U.S. in search of the "Cities of Gold"
1610	Spanish arrive in Santa Fe
1680	Pueblo Rebellion in New Mexico
1776	Dominquez-Escalante Expedition
1803	First Americans into Taos
1836	Texas Independence
1846–8	Mexican–American War; Treaty of Guadalupe Hidalgo
1848	Gold discovered in California
1849	Gold discovered in Colorado
1854	Gadsden purchase
1861–5	Civil War in the Southwest
1863–8	Navajo and Apache internment at Fort Sumner
1886	Geronimo surrenders

4 THE WINES OF SOUTHWEST U.S.A.

A word on climate change

It's worth noting that the subject of climate change is not widely addressed in each chapter of this book. It is indeed true that climate change has affected the region of the Southwest. The U.S. Environmental Protection Agency estimates that temperatures have increased by between 1 and 1.5°F (0.56–0.8°C) over the last century. The rate of groundwater depletion in the Southwest has increased severely over the past century, particularly in the past 20 years, as a result of increased demands on aquifers. But when it comes to grape growers and wine producers making adjustments in the vineyard and the cellar as a response to climate change, the reality is that each of these industries is still too young to have built up systems that would need to respond to the change. As each of these states is entering into their first few decades of defining their identities in the global wine sphere, significant historical data with which to compare growing seasons is lacking and is instead being written with each coming year. The most critical challenges that are arising from global warming in these regions include an increase in the severity of weather events from intense rainstorms with hail and wind, to substantial flooding, and more extended periods of drought. While many established regions in milder climates are re-evaluating viticultural techniques and even looking to different grape variety selections, there may be a silver lining for the already hot, dry climates of the American Southwest. The trend over the past 10–15 years throughout the region has been to plant warm-climate varieties that already thrive in the heat and can better tolerate drought. With this in mind, it would seem that in this time of discovery for these new winegrowing industries, the region of the Southwest is uniquely poised to withstand the impending changes brought on by the effects of climate change.

HOW TO USE THIS BOOK

The book is divided into four parts, one for each featured state. It begins with New Mexico, which is the first place in the United States to have planted *Vitis vinifera*—more than a century before California. Texas follows as it arguably has the second-longest history of viticulture. Arizona is third, though when it comes to the quality of wine, in many regards, it could well be considered first. Colorado is fourth, if only because its location in the northern part of the Southwest gives it a different landscape altogether.

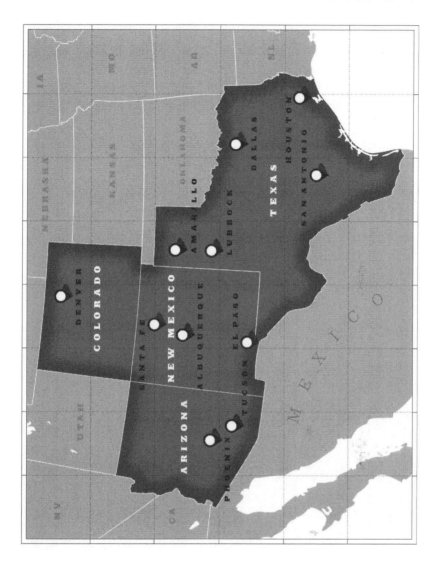

The history, regions, climate, geography, successes, and challenges of each are covered in as much detail as possible. In addition, each state includes a short but serviceable guide to a few places to dine, stay, and see should you find yourself in any of these areas. Throughout each section of the book are informative sidebars that may have less to do with wine, but more to do with key elements of the region that help paint a complete picture of the character and culture of the Southwest.

Each section includes a selection of producer profiles. Here it is important to note that in the whole of the Southwest region combined,

6 THE WINES OF SOUTHWEST U.S.A.

there are roughly 700 wineries. But as this quadrant of the country is considered to be "emerging", or perhaps more appropriately, "becoming", it was necessary to include just some of these producers. In each state the listing represents about 10–15 percent of the overall number of wineries. These are producers who are either leading the way in taking the industry to the next necessary level or who have played a significant role in the history of each state as its wine regions have evolved. While there are many others who could have been included, the hope is that those represented in this book give an accurate depiction of the future for these burgeoning regions. In numerous ways, this selection serves as a blueprint for the many great things that lie ahead.

Doug Frost: a Master of Wine looks at the American Southwest

Throughout the world of regional American wine, Doug Frost has become a household name. He is one of only four individuals in the world to hold the titles of Master of Wine (MW) and Master Sommelier (MS) concurrently. His willingness to give wines from America's emerging regions not only a chance, but also a resounding endorsement, has made him one of the greatest champions for these burgeoning industries—and the Southwest states of Texas, New Mexico, Arizona, and Colorado are no exception.

His confidence in the potential for wine regions around the country led him to establish the Jefferson Cup Invitational Wine Competition, a twenty-year-old national competition, which evaluates wine entries from around the country based on style and grape variety rather than region.

"I've been following these less familiar regions alongside California, Oregon, and Washington for the past four decades and early on I realized that there are really smart people who were fighting these uphill battles in these different states with very unexpected successes," says Frost. "That made me realize that I needed to be less closed-minded to what's happening out there."

Frost references a 1975 Chateau Ste Michelle Cabernet Sauvignon from Cold Creek Vineyards as the wine that changed his view about what Washington could do at a time when no one was taking the state's winemaking efforts seriously.

Today, Frost has been quoted as saying that "If I could wave a magic wand, I would make every restaurant in America start buying from local wineries. At least make sure they know what's going on."

Below are a few of Doug's thoughts on the wines of the American Southwest.

Jessica Dupuy: A little over 15 years ago, you helped start the "Drink Local" movement for American wine with *Washington Post* wine columnist, Dave McIntyre, and wine blogger Jeff Siegel. What was the impetus behind that?

Doug Frost: It was around the same time as the Farm to Table movement was trending, and a lot of us in the wine industry were wondering why the hell everyone was only talking about buying local produce, meat, and vegetables, but not wine. How are those local things "cool", but local wine is not? We tried to hammer at that question through media and events, including the Drink Local Wine conference, which was short-lived, but not without a good effort.

JD: You also helped to launch the Jefferson Cup, which puts a focus on wines from across America. Has this improved the perception of wines from American emerging regions?

DF: The original idea was it would be a wine competition for Virginia wines in the interest of helping to propel that industry forward. But to me, the idea of putting wines next to each other from each state was a better way to help each region learn how to improve. So, we made it an American-only wine competition. The wines are tasted in a blind setting according to style or grape variety but intermingled among the various regions. So you may taste a Virginia Cabernet Franc next to a Colorado Cabernet Franc, but you're simply evaluating the wine for its virtues based on that grape variety. The idea was to give everyone a level playing field, and after 20 years, I think we've made some small but measurable impact.

JD: Do you see promise in the Southwest region of America?

DF: California represents the vast majority of wine sold in America, [but] that is as much due to history as it is to quality. Certainly, in the 1950s, there was no guarantee that California would become dominant. I've had enough wines from other areas in North America to believe that other regions can challenge the state, particularly as the big, high-octane California style loses its allure for many younger consumers.

JD: What have you seen as some of the success factors for wines in the Southwest?

DF: Definitely Rhône varieties have shown themselves to be adapted to the terroir, but other grape varieties have shown well such as Aglianico, Tempranillo, Sangiovese, Vermentino, and Picpoul.

JD: What do you see as some of the region's greatest challenges?

DF: There aren't enough high-quality vineyards in the region at present; that needs to change. Certainly, the market outside of the region is not yet pursuing wines from the Southwestern states. That too is a challenge, but one that I believe can be slowly overcome, one good wine at a time.

JD: You are a judge for several regional and American wine competitions in addition to the Jefferson Cup. What are some of your takeaways regarding the regions of the Southwest, specifically Colorado, Texas, New Mexico, and Arizona?

DF: I like Colorado wines a lot. They really have some great producers who are constantly battling frost and other weather catastrophes. When they can avoid that, they make really solid wines.

Having lived in Fort Worth, I feel I've always had a conflicted relationship with the great state of Texas. I think for a while, they were too easy-going about it all. Just because you can grow Chardonnay, doesn't mean you should. But a lot of Texas producers like Kim McPherson have made strides with warm-climate varieties like Tempranillo, Sangiovese, and Rhône varieties, and these are the wines I look forward to tasting.

In the case of New Mexico, Gruet has long been the most recognizable name. They were the first to really stand up to California and say, "I can do this, too!" And be really successful at it. And despite the fact that they make very little New Mexico-grown wine today, you can't ignore the fact that what they have done has not only paved the way for New Mexico wine, but all of the Southwest.

In Arizona, there's just so much great wine. If I were going to have to pick one state in the Southwest that really has its act together in terms of making great wine across the industry, it would be Arizona. In the same way as Gruet, you can't underestimate the importance of Maynard James Keenan. He has not only used his fame to help create publicity for his own wines, but he's also set a benchmark of standards. When you consider the hard work that producers like Kent Callaghan and Todd Bostock have put into making great wine, their efforts have only been helped by Keenan, who has been a great megaphone for the industry.

PART 1
NEW MEXICO—WHERE THE SOUTHWEST BEGINS

1

HISTORY

While most American history is first associated with British colonization along the east coast of the country at the beginning of the seventeenth century, New Mexico was first colonized by the Spanish. Second to Florida, New Mexico is one of the oldest state names in the country. It was proclaimed *Nuevo Méjico* in 1565 by Spanish explorer Francisco de Ibarra in his expeditions north of Mexico.

New Mexico also has the distinction of having the longest history of wine production in the United States. As early as 1629, vineyard plantings were recorded among the Franciscan friars who established missionary settlements northward along the Rio Grande. The missions were the result of Spanish exploration by Don Juan de Oñate, the son of a wealthy silver baron in Zacatecas, Mexico in the late sixteenth century, who charted a new route up through the Chihuahuan Desert which later became known as the El Camino Real, or Royal Road, and would become the major trading route from Mexico to the New World. His initial quest was in search of gold for the Spanish monarchy, but he was also charged with establishing a settlement for the Christian colonization of the native Puebloan tribes of the region. Along with him, he brought colonists, Franciscan monks, soldiers, and cattle to establish a foothold in the area. In 1598, he established the settlement of San Juan de los Cabelleros, taking over the existing Ohkay Owingeh Pueblo, which was located 30 miles (48 kilometers) north of present-day Santa Fe. Although the area did provide excellent land for livestock grazing and relatively fertile soils for crop farming, because of the lack of gold or any other precious metals the endeavor ended up costing the Spanish crown more than it cared to continue investing. In 1608,

12 THE WINES OF SOUTHWEST U.S.A.

with the settlement in danger of abandonment, Friar Lázaro Ximénez traveled to Mexico as an ambassador for Oñate to plead for the continuation of the establishment.

The missions achieved some success in their quest to convert native inhabitants of the region—primarily Pueblo, Apache, and Navajo—but had a major obstacle in the presentation of sacramental wine. Vineyard establishment in the New World had begun as early as 1521, in Mexico's Valle de Parras, when noted Spanish explorer Hernán Cortés required that on the land they were granted, each Spanish settler plant 1,000 vines for every 100 natives in service to them. The effort was geared to producing sufficient wine for everyday life within the settlement. More than seven decades later, Spain's King Philip II noticed a steep cut in the exportation of Spanish wine to the New World. At the time, grape production accounted for nearly a quarter of Spain's foreign revenue. In an attempt to control revenues from the sale of wine and protect the Spanish agricultural industry at all costs, Spain passed a law in 1595 that prohibited grapevine growing in the New World. Instead, the missions were required to await shipments of wine from Spain via Mexico. This could entail a wait of as much as three years, assuming shipments weren't fated to sink to the bottom of the ocean, due to frequent shipwrecks in the Gulf of Mexico. What little wine did make it to the outpost at New Mexico was often maderized and spoiled from the long, oppressively hot transport through the Chihuahuan Desert.[1]

As a result, and in defiance of the Spanish order, Jesuit priests of Parras, Mexico began planting grapevines at the turn of the sixteenth century and making wine for sacramental use. The idea—and most likely a handful of vine cuttings—made its way to the Piro Pueblo of Senecú, near what is now the city of Socorro, where Franciscan monk Fray Garcia de Zúniga and Capuchín monk Antonio de Arteaga also planted grapes at the mission church, and through the churches of the Rio Grande in 1629. But these vineyards were not long-lived. Throughout the seventeenth century, a combination of harsh winters, massive flooding along the Rio Grande, and frequent skirmishes between the Spanish settlers and the warring native Apache tribe all but destroyed most of the vineyards as well as the entirety of the Senecú Pueblo.

---

1 Historians have also noted possible contamination of the wine due to heavy lead content in the ceramic glaze used for the clay pots in which the wine was stored.

The Pueblos of New Mexico

The Pueblo Period is believed by most archeologists to have begun more than 600 years ago when nomadic hunter-gatherers began settling and raising crops. By the fourteenth century, many of the major river valleys in New Mexico were supporting large Pueblo communities that depended on agriculture to maintain their growing populations. The Puebloan culture has been characterized as one rich in tradition, dependent on trade and family ties. Historically, the Puebloans resided throughout the Four Corners area of New Mexico, Arizona, Utah, and Colorado.

When the Spanish arrived in the mid-sixteenth century, it's possible that there were more than 100 pueblos located along the Rio Grande valley, from northern Taos Pueblo to southern Isleta Pueblo. Spanish colonization of these areas brought a slow progression of tension and unrest. In 1680, the Pueblo Revolt forced the Europeans to leave New Mexico until 1692 when the Spanish reoccupied the middle Rio Grande valley. Since that time, Spanish and Pueblo cultures have merged, and blended spirituality and tradition.

The Pueblo people relied heavily on the Rio Grande valley for their water supply, developing a primitive irrigation system of ditches throughout the region using a series of main *acequias* (shared irrigation ditches), off which individual trenches were designed. The Pueblo *acequia* system is one of the oldest water systems recorded in North America.

The word "pueblo" is Spanish for "village", and was used by Spanish explorers to describe the people of this pre-Columbian tribe living throughout the middle Rio Grande valley. But the term was also used to describe the residential structures made up of a series of multi-story flat-roof houses constructed from stone, wood, and adobe brick, which is a mixture of clay, water, and straw. This architectural adobe-style has been continued in today's New Mexico structures. In fact, it's a standard feature throughout the historic districts of Albuquerque, Santa Fe, and Taos.

Today, only 19 pueblos remain in New Mexico, each with its own government but sharing a common prehistory and culture. The Acoma "Sky City" Pueblo, which sits atop a large, flat mesa about 60 miles (96.5 kilometers) west of Albuquerque, is the oldest continuously inhabited community in North America. It is one of many pueblos open to the public to visit today, and is one of the best ways to experience the richness of this culture. There are several near Albuquerque as well as Santa Fe and Taos.

It is important to note that all pueblos are sovereign nations, administered by tribal governments. Each has its own set of tribal customs. For instance, although most pueblo communities are open to the public during daylight hours, the homes are private and should not be entered without an invitation. *Kivas*, or underground ceremonial rooms for rites and meetings, and graveyards are not to be entered by non-Pueblo people. Many of the walls and other structures within the pueblos are several hundred years old. It is common for visitors to be prohibited from touching the structural walls. Pueblo dances on ceremonial days are religious ceremonies, not staged performances. Interacting or interfering with members of the tribe during these ceremonies is unwelcome, as is applauding after the dances.

Each pueblo has its own regulations regarding photography, video, artwork, and audio recording. Official permits for any photography are typically required upon entering the pueblo. A permit does not grant you the right to take photos or make sketches of an individual without asking permission first. Some pueblos forbid photos of *kivas*, cemeteries, and church interiors.

Having tired of the strict requirements of Spanish occupancy, many of the Puebloan tribes banded together to stage the Pueblo Revolt of 1680, which resulted in the killing of more than 400 Spaniards with an additional 2,500 driven south into Mexico and west to the village of El Paso in what is now the state of Texas. As a claim to victory, all traces of Spanish occupation were burned and buried, erasing almost all evidence and documentation of the history in the region during this time.

But Spanish conquest would resume just over a decade later when Don Diego José de Vargas Zapata Luján Ponce de Léon y Contreras, also known as Don Diego de Vargas, re-established Spanish settlements in 1692. As part of his mission he had been commanded to replant grapevines throughout the reclaimed region. In the following decades, wine production among the missions and private land holdings would flourish. In a diary entry from a 1766 expedition by Franciscan friars to report to the Spanish monarchy on the fortified military bases, or *presidios*, of New Mexico, engineer Nicolas de La Fora wrote of his surprise at the quality of the New World wines he encountered. He described them as being "In no way inferior to those of Spain." It's worth noting that the California wine industry began with the planting of the Mission grape variety in 1769, which is three years after this report was made, and nearly 140 years after the first vineyard plantings in New Mexico.

In 1804, wine was listed as one of the top three exports from New Mexico, along with wool and animal pelts. As the region progressed into the early nineteenth century, New Mexico, along with Arizona and Texas, remained under Spanish rule. This ended when Mexico won its independence in 1821. Little is recorded or known of the resilience of the mission vineyards. The establishment of the Santa Fe Trail by William Becknell in 1821 saw an influx of Americans into the Rio Grande valley; they were also encouraged by rumors of its grapes and wine production. As the center of trade between the eastern United States, Mexico, and Mexican California over the next few decades, wine of New Mexico was often purchased by travelers along the trail and commonly used as currency for trade. During this time, there are ample reports of vineyard plantings and wine and brandy production throughout the Rio Grande and Mesilla valleys.

With the annexation of New Mexico as a U.S. territory in 1848, talk of its potential as a wine-growing region began to circulate. According to U.S. attorney of the territory, William Davis, "No climate in the world is better adapted to the vine than the middle and southern portions of New Mexico, and if there was a convenient market to induce extensive cultivation of the grape, wine would soon become one of the staples of the country, which would be able to supply a large part of the demand in the United States."

Traffic through New Mexico increased with the beginnings of the California Gold Rush in 1849, introducing local wines to the droves of travelers along the Santa Fe Trail.

Indeed, by the middle of the century, vineyard plantings could be found from Las Cruces in the south, near El Paso, Texas, all the way through to the northern part of the state, near the Colorado border. In the following decades, plantings would increase significantly under the influences of Spanish Jesuit missions throughout the state. In 1868, New Mexico saw an influx of Italian Jesuit priests who had been expelled from Italy by Giuseppe Garibaldi and who brought with them Old-World winemaking skills. In addition, a push towards French winemaking traditions came under the direction of native Frenchmen; archbishop Jean-Baptiste Lamy was influential, and along with French vintner Joseph Tondre increased the quality of winemaking compared to the more rudimentary styles that had been produced to date. By 1880, the state's wine production had risen to 980,000 gallons (3.7 million liters), and with more than 3,150 acres (1,275 hectares) under vine, more than twice that of New York, it had earned the ranking of fifth in the nation for wine production.

The grape mission

It is unclear exactly which grape variety was planted by the Franciscan monks in 1629, since there are no remains of the original vineyard plantings. However, many believe that it was the Mission grape. According to research from the Centro Nacional de Biotecnologica in Madrid, it was the earliest *Vitis vinifera* grown in the western hemisphere. It is also believed to be the first grape variety planted in the Valle de Parres, Mexico, which is likely where cuttings for the early New Mexico planting originated. Of course, it wasn't known in Spain as the Mission grape. In its native country, it was Listan Negro, or Listan Prieto, a native Spanish grape that was known for its thick skin and sturdy, hardy trunks. It proved resilient in transfer to the New World as well as productive in the dry, mineral-rich soils of New Mexico. The variety has been genetically linked to Listan Blanco, or Palomino Blanco, the primary grape in Sherry production.

Mission grapes form large, loose clusters that allow air to move through the grapes, greatly reducing the risk of rot and mold. The result is a cluster that can take more time to ripen on the vine. Depending on growing conditions, the vines can easily produce more than 10 tons per acre (22.42 metric tons per hectare) if yields are not managed. They were probably allowed to push for the higher quantities for the Spanish mission sacramental wines in the New World.

In terms of wine production, there's a reason Mission isn't topping the list of the wine world's most planted grapes. Though it has rather thick skins, the fruit rarely develops much concentration either in color or phenolic content. Wines tend to lack structure, depth, and acidity. Though it served its purpose for sacramental wine, it generally falls short in expectation compared to the international varieties that are planted throughout the western hemisphere today.

Today, Listan Negro is rarely seen in Spain, aside from some plantings on the Canary Islands. But it seems to have found a home in the Americas. In South America, in particular, it is grown for bulk wine production and in recent years for serious table wines. Argentina grows it under the name of Criolla, Peru calls it Negra Criolla, in Bolivia it is Missionara, and in Chile, it's País. While not widely planted in the United States, you can find it in California's Santa Barbara regions as well as further north near Sacramento.

You can also find it at Tularosa Vineyards Winery in the small town of Tularosa in Southern New Mexico, just east of the Rio Grande. With elevations ranging between 4,500 and 6,800 feet (1,372 and 2,072 meters), the Tularosa Basin, where the town is situated, first saw vineyards in the 1860s from Mission vine cuttings brought by Spanish settlers.

According to an article in the *Alamogordo News* in September 1932, describing the town of Tularosa in 1870:

Grapes were planted, and wine was made to supply the local requirements. The principal product was a native sour wine locally known as "cerque" and was freely used by many residents. The usual method of manufacture was very primitive and crude, and in present times would be condemned as unsanitary and unfit for human consumption. The grapes were emptied into rawhide containers; sack like affairs (tinas) suspended on forked posts, which in turn carried square frames of poles to which the rims of the tramping sacks were lashed. Men in their bare feet tramped them to a pulpy mass, sinking to their knees in the process. The "pummies" were then skimmed from the surface as they rose to the top of the fermenting wine. After 10 days without disturbing, the sediment collected in the bottom, and cool fermentation was allowed to complete. The wines were deemed ready 30 days from the second draining. Some wines produced were a well-flavored stimulating beverage, while others made a vinegar-like product that intoxicated and sickened with a nauseating flavor. The most palatable was, of course, in greater demand and brought a better price, but the inferior stuff was still consumed at a profit to the producers.

Nearly a century later, Tularosa Vineyards Winery was founded in the late 1980s by New Yorker David Wickham and his wife, Teresita. Following a tour of duty in the U.S. Air Force, the Wickhams settled in the quaint New Mexico small town and stumbled into the hobby of grape growing and winemaking. This endeavor would later lead to the planting of 10 acres (4 hectares) of vineyard in 1985 and the opening of their winery in 1989. In 1989, Wickham planted a small vineyard of Mission grapes from the cuttings of a vine he found in the backyard of a man named Ben Chavez in La Luz, New Mexico. The vine was estimated to be a little more than 100 years old. With a trunk diameter of approximately 30 inches (76 centimeters), the single vine is said to have yielded more than 300 pounds (136 kilograms) of grapes in different years. To preserve the historical relevance of the grape within New Mexico, Wickham has cultivated an acre of Mission vines from which he produces in an off-dry style.

In the following decades, Wickham served as a mentor to other pioneering farmers who were interested in trying their hand at grape growing, buying their yields for his expanding wine production. In 1995 his son took on the role of winemaker, adding Tularosa Vineyards Winery to the growing list of multi-generational family wineries in New Mexico. Today, the winery still operates at a small capacity but produces more than 3,500 cases annually.

18   THE WINES OF SOUTHWEST U.S.A.

Almost all growth came to an abrupt halt in the early twentieth century with the rise of the temperance movement. Making its way throughout the southwest part of the country, the movement took a stronghold in New Mexico with the Woman's Christian Temperance Union, the American Temperance Society, and the Anti-Saloon League all gaining influence in Albuquerque and Santa Fe. Growing support for these organizations led the state to adopt its own statewide Prohibition laws in 1917, two years before the United States government created the Eighteenth Amendment, declaring the production, transport, and sale of intoxicating liquors illegal. While wine was still produced in small amounts for medicinal and sacramental purposes, almost all wine production ceased for more than a decade, and vineyards had to be plowed and replanted to fruit orchards or other crops.

Although Prohibition was repealed in 1933 with the Twenty-First Amendment, New Mexico vineyard development had little chance of recovery, with the Rio Grande valley falling victim to a devastating 100-year flood in 1943.

Burgeoning hope returned to the New Mexican wine industry in 1977 with the opening of La Viña Winery in La Union, on the other side of the Texas border from El Paso. La Chiripada Winery opened soon after in Dixon, planting cold-hardy French-hybrid vines at its cooler, high-elevation site. In 1978, the industry received a boost when a government-sponsored study deemed New Mexico's soil and climate favorable to these newly created hybrid grape varieties. As a result, wineries slowly began production, and tasting rooms began to open their doors.

In the 1980s, a large number of Italian, French, and German immigrants brought their viticultural expertise. After all, New Mexico was a place where an acre of land sold for pennies on the dollar compared to California. Its growing conditions were said to have much in common with great Old-World wine regions such as the southern Rhône valley, Central Italy, and Southern Spain: dry, with rocky, calcium-rich soil. Between 1982 and 1983, 2,200 acres (890 hectares) of vineyards were planted around Las Cruces alone.

In 1981, Hervé Lescombes, a fifth-generation winemaker, left the successful Domaine de Perignon winery in Burgundy, France, to plant vines in New Mexico. Lescombes' endeavor has steadily grown over the past three decades with award-winning wines made from iconic French grape varieties such as Cabernet Franc, Mourvèdre, Cabernet

Sauvignon, and Merlot. The next generation of Lescombes, Emmanuel and Florent, now manage Southwest Wines, a 500,000-gallon (18,927 hectoliter) capacity winery in Deming, producing multiple labels under the Lescombes name.

Among the Italians was Paolo D'Andrea, a fourth-generation wine grower from Friuli. D'Andrea arrived in Deming, New Mexico, in 1986 to manage the state's largest vineyard, in the Mimbres Valley. By 2001, he and his wife, Sylvia, had founded their own winery, Luna Rossa, as well as a nursery that continues to supply grapevines throughout the Southwest today.

New Mexico received its first sparkling-wine house when Frenchman Gilbert Gruet, of Gruet et Fils winery in Champagne, soon after Lescombes, began purchasing land for vineyard development in Engle near the Elephant Butte Reservoir in 1984. Nearly four decades later, Gruet is one of the largest sparkling wine producers in the country, with a wide distribution in 48 states, and is one of the most nationally-recognized sparkling wine producers in the country. Today, Gilbert's son, Laurent, is head winemaker, while his daughter, Nathalie, is president of the company.

In 1984, Bernd Maier brought his family from Baden, Germany, to Engle, New Mexico, and began planting vineyards and experimenting with different varieties that could be a fit for the region. His extensive knowledge of vineyard management and keen intuition on grape selection eventually made him the number one contact new wineries sought for advice. In 2004, the state legislature provided initial funding for an agricultural extension service, which serves as an information resource and a guide to farmers, ranchers, and community groups. In 2006 Maier was asked to become the Extension Viticulture Specialist for the state, going to work installing a climate station network throughout the wine-growing regions. Since then, he and his son, Benjamin, have served as consultants for vineyard planting, irrigation, and frost protection. The two also own and operate Amaro Winery in Las Cruces. In addition, he has taught many growers throughout the state about the importance of the trellising systems, canopy management, and grape varieties that are best suited to New Mexico's climate.

Early efforts by small and new producers to gain entry into the market throughout the 1980s and 1990s were met with challenges well beyond climate and growing conditions. Competition from California and Europe loomed large over New Mexico's wine sales making it nearly impossible for producers to create long-term goals. But fate began to smile on New Mexico wine with the help of New Mexico State University's School of Agriculture. Located in Las Cruces, the school has developed a program to advance viticultural and winemaking education, including a new experimental vineyard at the university's Fabian Garcia Research Center, where 34 grape varieties are planted.

Today much of the success and growth in the New Mexico wine industry can be attributed to the adoption of the adage "there's strength in numbers". To solidify the general interests and goals of producers and wine growers within the state, the New Mexico Wine Growers Association was launched in 1991 as a nonprofit organization. With 55 wineries spread out across the state, membership of the organization has allowed for a greater message about positive growth within the industry, most of which focuses on the communal nature of the wine world. In 2016, the organization launched a "Viva Vino" campaign, a universal slogan used throughout the wine regions to celebrate the multicultural heritage behind New Mexico's wine.

2

CLIMATE, REGIONS, AND CHALLENGES

CLIMATE

New Mexico is renowned for a particular beauty definitive of the American Southwest. With its sweeping landscapes of serene desert terrain, historic adobe villages, and bewitching pastel-hued sunsets, it's no wonder New Mexico is referred to as the Land of Enchantment. But for all its austere beauty, it is also rugged, desolate, and at times, unforgiving. With arid desert landscapes riddled with scrub-brush and parched, cracked earth, it's hard to imagine why anyone would envisage it as a location for verdant vineyard plantings. But with the sandy, alluvial soils that grace the expanse of the fertile Rio Grande river basin, along with the arid environment and high-elevation sites, it turns out New Mexico is uniquely primed for vineyards, a fact that was discovered more out of necessity than by design. New Mexico may be the fifth-largest state in the country by sheer geographic size—121,589 square miles (314,914 square kilometers) to be exact—but it's also one of the least densely populated states. With only about 2 million people living in this sunny state, its population is half that of the city of Los Angeles.

New Mexico's wine industry doesn't simply owe its current uptick to its fortunate history as the veritable birthplace of American wine. The fact that it has favorable growing conditions in both its southern and northern regions is a critical factor in its overall success.

With a sweeping landscape of high plateaus, numerous mountain ranges, canyons, valleys, and meandering dry *arroyos*, elevations

21

throughout the state vary from 2,817 to 13,161 feet (859 to 4,011 meters) at its highest mountain peak, Wheeler Peak.

New Mexico generally has an arid or semi-arid continental climate with regular clear skies and low relative humidity. Temperatures vary greatly depending on season and elevation. During the warmer growing season in the summer, daytime temperatures can exceed 100°F (38°C) at elevations lower than 5,000 feet (1,524 meters). However, the average monthly maximum temperature during the warmest month of the year, July, is just over 90°F (32°C) in lower elevations, and in the upper 70s (around 25°C) at higher elevations. In the lower hills of the southern valleys of New Mexico, the frost-free season averages about 200 days. The dry climate assists in deterring the rot, mildew, and vineyard pests commonly found in more humid regions.

Statewide, average annual rainfall ranges from less than 10 inches (250 millimeters) over much of the southern desert and the Rio Grande and San Juan valleys to more than 20 inches (500 millimeters) at higher elevations in the state. The southeasterly circulation of the Gulf of Mexico, more than 500 miles (804.5 kilometers) to the south, generates brief but frequently heavy thunderstorms during July and August, leading to about 35 percent of the year's total rainfall. Cold, dry winters are often mitigated by irrigation during colder months to encourage deeper rooting, and longer canes to combat frost.

GEOLOGY

New Mexico's landscape is defined by both the Basin and Range Province and the Rio Grande Basin, shaped by the Rio Grande.

Spanning almost the entirety of the state is a large portion of the Basin and Range Province, an expansive physiographic region that extends over much of the inland western United States and northwestern Mexico. The area is the result of a tectonic extension from more than 17 million years ago featuring an assortment of abrupt elevation swings, narrow mountain chains with multiple fault lines, and large swaths of flat, arid valley. Among the many ranges included in the province are the Snake and Panamint ranges in California, and the White Mountains in Arizona. Within the state is the Sandia Mountains range, which stretches 17 miles (27 kilometers) from north to south just east of Albuquerque. These ranges extend from the earth as narrow, parallel elevations that snake from north to south throughout the region.

Geologic point of interest: White Sands, New Mexico

One of New Mexico's defining geologic features, White Sands National Park is among the world's great natural wonders. Rising from the Tularosa Basin in the south-central part of the state is a series of glistening wave-like dunes made of gypsum sand. The 275-square miles (712 square kilometers) of desert represent the world's largest gypsum dune field.

The formation is the result of massive gypsum deposits left from shallow seas that covered the area during the Permian period. Subsequent tectonic activity lifted areas of the gypsum-rich seabed to form part of the San Andres and Sacramento Mountains, which surround the area. Over time, rain dissolved the gypsum in the mountains, and rivers carried it to the Tularosa Basin, which has no outlet to the sea. The trapped water sank into the ground or formed shallow pools that subsequently dried out, leaving the gypsum on the surface in a crystalline form called selenite.

During the last ice age, there was a large lake covering much of the basin. When it dried out, a large flat area of selenite crystals remained, which is known as the Alkali Flat. Around this area, the ground consists of selenite crystals that measure up to three feet (a meter) long. Over time, weathering and erosion break the crystals into sand-size grains that are carried away by the prevailing winds from the southwest, forming the white dunes. The dunes constantly change shape and slowly move downwind. Since gypsum is water-soluble, the sand that composes the dunes often dissolves and cements together after rain, forming a layer of sand that is more solid, which increases the wind resistance of the dunes.

The dunes were open land for many years, with mining companies seeking to take over the formation throughout the late nineteenth and early twentieth centuries. By the 1930s, the area had gained enough support to become designated as a National Park, taking the name White Sands National Monument in 1933. In early 1942, only a few months after the attack on Pearl Harbor, President Franklin D. Roosevelt designated the 1.2 million-acre (0.5 million hectare) Alamogordo Bombing and Gunnery Range adjacent to White Sands for military practice. The military space is now the Holloman Air Force Base, and the range, which is still used for missile testing, is the White Sands Missile Range. Both military areas still operate around the park boundaries, which has benefitted the entire region by ensuring the lack of development on the surrounding lands.

The Rio Grande and its resulting river valley have played a significant role in the development of the state's wine-growing regions. Vacillating between the fourth and fifth longest river in North America due to its occasional course changes, the roughly 1,900-mile (3,058-kilometer) river begins in a small canyon of the Colorado Mountains. It flows south, cutting the Rio Grande Gorge and the White Rock Canyon of northern New Mexico before flowing into the open expanse of the Basin and Range Province and meandering further south to form the border between Texas and Mexico, ending its journey in the Gulf of Mexico through the Mexican Chihuahua desert.

In New Mexico, the Middle Rio Grande Basin, also known as the Albuquerque Basin, is one of the largest structural basins in the Rio Grande rift. Though the climate of this geographic feature is considered semi-arid today, there is evidence to suggest that around 12,000 years ago the region was wetter and more fertile. The basin relies heavily on the Rio Grande for irrigation and drinking water. Covering about 3,100 square miles (8,029 square kilometers), it is bound by the Sandia and Manzano mountains to the east, the Jemez Mountains to the north, the Rio Puerco on the west, and the Socorro Basin to the south.

Geological characteristics of the area include alluvial fans from mountains that form the eastern boundary of the basin as well as isolated volcanoes and mountains to the west depositing basalt and andesite rock along the west side of the basin. Alluvial sediment can be found from the adjacent highlands as well as fluvial sediments from southern Colorado and northern New Mexico.

New Mexico's soils are characterized as rocky and sandy, providing excellent drainage, which helps to safeguard against rot deep within the roots. The Rio Grande valley is located in the lower plane of the state and is made up entirely of alluvial soils containing loose river sediments of clay and silt. The groundwater throughout the region generally has high salinity, which can have adverse effects on vines.

REGIONS

The state is home to three American Viticultural Areas (AVAs), including the Middle Rio Grande Valley, the Mimbres Valley, and the Mesilla Valley, which spills into Texas along the eastern border. In total, New Mexico has about 1,500 vineyard acres (607 hectares) planted, the majority of which are in the Mimbres Valley AVA.

Mimbres Valley

Located in the southwestern part of the state, the Mimbres Valley AVA was established in 1985. Spanning more than 636,000 acres (257,380 hectares), it is New Mexico's largest viticultural area, both in geological boundaries and in overall plantings—accounting for more than half of the estimated 1,500 acres (607 hectares) planted in the state.

The majority of the vineyard plantings in the entire state are located near Deming, in the southwestern corner, and fed by the watershed of the Mimbres River between Deming and Silver City. Soils are predominantly alluvial from river deposits, including some clay and loam. Here, elevations are around 4,300 feet (1,311 meters) but can rise to 6,000 feet (1,829 meters) in other parts of the region. Considered to have a continental climate, the area is marked by hot, sunny summer days, cool nights during the growing season, and cold winters. The average rainfall for the Mimbres Valley is 9 inches (229 millimeters), thus making drip irrigation a necessity for vineyard growers.

Many have compared the growing conditions of the Mimbres Valley AVA to that of Argentina's Mendoza region—minus the Andes mountains, of course. The largest and oldest vineyards are located in this region, and other wineries throughout the state commonly purchase grapes grown in the Mimbres Valley. New Mexico's largest vineyard, at more than 200 acres (80.9 hectares), is planted on the border of Hidalgo and Luna counties and is owned by the Lescombes family, pioneers of the state's modern wine industry. The vineyard is planted with more than two dozen varieties, including Chardonnay, Cabernet Franc, Petit Verdot, and Mourvèdre. The region is also home to the state's other largest wine grower, Paolo D'Andrea of Luna Rossa, who farms more than 300 acres (121.4 hectares). Cabernet Sauvignon, Syrah, Tempranillo, and Italian varieties such as Dolcetto, Sangiovese, and Aglianico are the predominant grapes grown in this region.

Mesilla Valley

When explorer Don Juan de Oñate came to this area at the turn of the sixteenth century, he named a local village Trenquel de la Mesilla, meaning "little table", referring to the small plateau on which the town sat. The plateau became the region's defining landmark, and would later lend itself to the AVA name. Considered more of a "partial" AVA because of its geographic overlap into the western part of Texas, the Mesilla Valley is the oldest AVA in the state. It extends along the edges of the Rio Grande down through Las Cruces and Anthony in New Mexico, with a small portion spilling into Texas at El Paso.

Established in 1985, the region covers about 280,000 acres (113,312 hectares), but with less than 50 acres (20.2 hectares) planted, it is New Mexico's smallest AVA. Etched into the landscape as part of the Rio Grande flood plain, the soils of this region are primarily alluvial sandy loam, clay, and sedimentary deposits from the nearby Organ and Franklin Mountain ranges. Elevations range between 3,700 feet (1,128 meters) along the Rio Grande to 9,012 feet (2,747 meters) in the peaks of the Organ Mountains. However, most vineyard plantings in the Las Cruces area sit at around 4,300 feet (1,311 meters). The region averages about 340 days of sunshine and sees temperatures during the summer season averaging about 96°F (36°C), falling to 28°F (-2°C) in the middle of winter. The average rainfall for the Mesilla Valley is only about 10 inches (250 millimeters), the majority of which falls in the higher elevations of the Franklin Mountains to the east.

Notable producers of the Mesilla Valley include Amaro Winery, Luna Rossa Winery (tasting room), La Viña Winery, and Lescombes Winery & Bistro (tasting room). Red grape plantings include Cabernet Sauvignon, Zinfandel, Syrah, and Tempranillo, while white grapes include Viognier and Sauvignon Blanc.

Middle Rio Grande Valley

The Middle Rio Grande Valley AVA was officially delimited in 1998, becoming the last of the three AVAs in New Mexico. As the name implies, this AVA is situated along the mid-section of the Rio Grande, in central New Mexico, from Santa Fe to just south of Albuquerque. The AVA spans 278,400 acres (112,664 hectares) with many vineyards planted at elevations ranging between 4,500 feet and 6,500 feet (1,372–1,981 meters). With intense sun exposure, along with hot days, and cool nights during the growing season, grapes can retain a desirable amount of acidity. The soils are primarily well-drained sand and loam. Winters are often very cold, and the small amount of rainfall makes irrigation a necessity. However, the dry climate is beneficial for keeping diseases such as powdery mildew and fungal vine disease at bay.

As with most of the state, the climate in the Middle Rio Grande Valley is semi-arid. Average annual temperatures are around 55°F (13°C) with average temperatures in the winter around 34°F (1°C), climbing to 75°F (24°C) in July. Average yearly rainfall depends on elevation, ranging from 7.5 inches (191 millimeters) about 35 miles (56 kilometers) south of Albuquerque in Belen, to 30 inches (762 millimeters) at Sandia Crest peak in the Sandia Mountains overlooking Albuquerque. Thunderstorms are common throughout the summer and winter seasons, though steady rainfall is unpredictable. The region has been known to sustain droughts lasting several years.

The AVA is home to nearly a dozen wineries, including sparkling wine house Gruet, which is located in Albuquerque. Other notable wineries or tasting rooms in the region include Casa Rondeña, Jaramillo Vineyards, Sheehan Family Winery, and Lescombes Winery & Bistro.

Northern New Mexico

Though not officially delimited as an AVA, it's worth noting the wine production taking place in the northern part of the state. About 45 miles (72 kilometers) north of Santa Fe, a handful of producers have tried their hand at growing in the colder climate, taking advantage of the higher elevations where vineyards are planted in sites from 5,500 to nearly 7,000 feet (1,676–2,134 meters). Here, growing conditions are significantly cooler during the summer months than in the regions further south, with temperatures only reaching 81°F (27°C). Still, the cooler night temperatures, which average 46°F (8°C), are desirable for cool-climate grapes such as Riesling, Chardonnay, Syrah, and Pinot Noir. Some producers, such as La Chiripada Vineyards, have turned to cold-hardy hybrids to account for the harsher growing conditions of colder, high-elevation sites. Notable producers, including La Chiripada and Vivác Winery, are working on having this area officially credited as the Northern Rio Grande Valley AVA.

GRAPES

Hand sorting Sangiovese at Vivác

New Mexico's unique geography affords a wide array of grape varieties, though considering its relative youth as a modern wine industry, it's fair to say that there is still much exploration ahead to determine which varieties perform best. As expected, Cabernet Sauvignon, Chardonnay, and Merlot are fairly common throughout the state. In the northern part of the state, cool-climate grapes such as Riesling, Gewürztraminer, Cabernet Franc, and Pinot Noir have been planted. In the south, particularly in the Mimbres Valley where the majority of the state's grapes are grown, there has been more experimentation with warm-climate grapes. Much of this has been through the influence of Luna Rossa's Paolo D'Andrea, and Hervé Lescombes from D.H. Lescombes. Leaning on their understanding of varieties from the warmer regions of their native Italy and France, the two began planting varieties such as Sangiovese, Dolcetto, Aglianico, Barbera, Montepulciano, Refosco, Viognier, Syrah, Mourvèdre, and Grenache as well as Tempranillo, and other Mediterranean varieties. Because the state of New Mexico does not require grape crop reports, there are no accurate statistics about what is planted, a challenge the New Mexico Wine and Grape Growers Association hopes to address in coming years.

CHALLENGES

Despite New Mexico's extensive wine history, there's still much to achieve as the industry looks ahead. From a vineyard standpoint, though suitable soils, desirable elevations, generally dry weather conditions, and cool nights all combine to present an excellent case for growing grapes here, there are several weather conditions throughout the year that keep producers on their toes.

Late spring frost, dramatic rainfall during harvest season, and raging winter snowstorms called "Goliaths", have all brought devastating effects to vineyards over the years. The substantial heat in the southern parts of the state, where the majority of the grapes are grown, presents a shorter growing season in which buds break sometime in early March, but may be harvested as early as mid-July. This fact does put New Mexico (and by extension its neighboring states, Texas and Arizona) at a disadvantage. High winds from March through to May, when the vines are experiencing fruit set, can produce shatter in some varieties. In addition, combating high heat, insects such as grape leafhoppers, as well as rabbits, foxes, javelinas (a native wild pig), and elk make a rigorous vineyard management strategy essential during the ripening season.

With a reputation for high pH, the struggle to maintain acidity in the grapes towards the end of the growing season is ongoing. While leaning into longer hang times for grapes to develop more phenolic ripeness, winemakers often have to acidulate in the winery to bring balance and vibrancy to the finished wines.

As with any burgeoning region, developing a cohesive identity is a hurdle to overcome. The fact that the state's best producers have started to embrace warm-climate grape varieties for the area has helped to define this identity. The New Mexico style tends away from the robust and ripe towards the more earthy and finessed flavors that are more commonly found in the Old World.

In recent years, producers have been more open to collaboration and sharing best practices than they were in decades past. With the help of the New Mexico Wine and Grape Growers Association (New Mexico Wine), the lines of communication and, more importantly, a cohesive message about the industry have helped strengthen its standing. New Mexico Wine assists members and the industry through education, legislation, and marketing. Under the tagline "Viva Vino", all New Mexico producers have come together to offer a more unified identity as a wine-growing region.

The organization has been able to help influence industry-positive legislation, including a "reciprocity law", which allows wineries to serve cider and beer (and vice versa) in their tasting rooms, thus capturing a wider audience and strengthening the identity of all three industries. New Mexico Wine has lobbied for the approval of private celebration permits, allowing wineries to host private events such as weddings. It has also lobbied for the approval of bottle service, which will enable bottles purchased in a winery or tasting room to be recorked and taken home. But there is more work to be done. New Mexico Wine has more recently been pushing for the addition of road markers along state highways to help inform and guide tourists to various wineries across the state.

The perpetual fight to find space on retail shelves and restaurant menus is a reality New Mexico wine shares with just about every other small producer in the world. But it does help that more and more consumers are willing to give New Mexico wine a shot, something that was not the case ten years ago.

3

NEW MEXICO PRODUCERS

Although the state's history spans more than five centuries, New Mexico is still a small wine industry compared to many of its compatriots. Over the years it has faced challenges from harsh growing conditions, Prohibition, and the difficulties of achieving a consistent quality. Despite its challenges, the families that pioneered the modern New Mexico wine era, Gruet, Lescombes, and D'Andrea, continue to help shape the industry today. As new, quality-focused producers such as Noisy Water and Vivác have joined the fold, the future is bright for this sun-baked state.

MESILLA VALLEY

La Viña Winery
La Union
lavinawinery.com

With the distinction of being the longest running winery in New Mexico, La Viña deserves credit for keeping the momentum going in the decades that followed its foundation. Located in a serene hacienda-style tasting room in the town of Anthony along N.M. Highway 28 (arguably New Mexico's best wine route), La Viña has maintained its status as a must-stop winery destination in New Mexico. Founded in the mid-1970s by Dr. Clarence Cooper, an associate physics professor at the University of El Paso, La Viña began as nothing more than a few experimental front-yard plantings. By 1977, Cooper had planted enough vines for his experiment to evolve into a commercially viable vineyard. Though he tried his hand at winemaking, it wasn't until the early 1990s

32 THE WINES OF SOUTHWEST U.S.A.

when Ken and Denise Stark took over ownership of the brand and eventually acquired the property that is now the winery's current home, a 40-acre (16-hectare) jalapeño farm, that wine production really took off.

In 2007, Chilean winemaker Guillermo Contador joined the La Viña team after answering an ad for a new winemaker. Originally from the small coastal town of Viña del Mar, Contador earned his degree in Agricultural Engineering and Oenology before working with notable wineries in Chile such as Santa Rita and Veramonte, as well as in Stellenbosch, South Africa at Boschendal, and in California at Jackson Family Farms, La Crema, and Clos du Bois. For him, something was intriguing about making wine in a lesser-known region of the United States. After all, his native country of Chile had only recently become a wine success story after years of lower-league struggle. The difference, according to Contador, is that he used to make wine in places where it made sense to make wine—near the ocean, with hospitable climates and favorable soils. Adapting to the different growing conditions at the 30 acres (12.1 hectares) of estate vineyards he works with at La Viña has been a challenge, but one to which he has readily risen.

The vineyard now grows 25 different varieties, including everything from Cabernet Sauvignon, Merlot, Sangiovese, and Dolcetto to Chardonnay and Semillon, all of which were planted before Contador arrived. Such a wide selection of grapes has granted him a blank canvas for creativity to make well-balanced, dry blends and single-variety releases as well as cater to the general local demand for sweeter-style wines. Among his favorite varieties to work with are Primitivo for its structure and Black Muscat for one of his sweet wine offerings. And as a nod to his coastal Chilean roots, Sauvignon Blanc is a personal favorite. Although he finds it much harder to work with in the warmer climate, he still manages to pull off a beautiful, nuanced wine each vintage.

MIMBRES VALLEY

D.H. Lescombes Family Vineyards
Deming and locations throughout the state
www.southwestwines.com

A true pioneer of the modern New Mexico wine industry, Hervé Lescombes is a sixth-generation French winemaker who, like his fellow Old-World compatriots, Paolo D'Andrea, Gilbert Gruet, and Bernd Maier, saw opportunity in the state's potential. Having spent his early career managing Domaine de Perignon in Burgundy, his affection for New Mexico stemmed from its unfettered freedom to grow grapes without the encumbering restrictions of French winemaking. In many ways the landscape was familiar, bearing a striking resemblance to parts of Algeria, the country of his birth. In 1981 he planted his Lordsburg Vineyard about 47 miles (75.5 kilometers) west of Deming in the Mimbres Valley. The 200-acre (80.9-hectare) vineyard is second in size to the New Mexico Vineyards managed by Paolo D'Andrea and sits at an elevation of about 4,500 feet (1,372 meters). At this altitude, the diurnal shift can swing day-to-night temperatures by 30°F (16.7°C) or more during the ripening season. Cabernet Sauvignon, Chardonnay, Merlot, Chenin Blanc, Syrah, and Mourvèdre are a few of the site's shining stars. All grapes are machine-harvested and are trucked east to the Lescombes winery in Deming.

Today, Hervé, who still spends time in the vineyard, has passed the day-to-day business on to his sons Emmanuel, who serves as the viticulturist of the family, and Florent, who manages the winery operations. Having steadily grown in both quality and size over the past three decades, the Lescombes production consists of more than 50,000 cases annually, all of which are distributed within the state. In addition to its local tasting room at the winery in Deming, there are three D.H. Lescombes Winery & Bistro locations in Albuquerque, Las Cruces, and Farmington, as well as the Hervé Wine Bar in Santa Fe.

Luna Rossa Winery
Deming
www.lunarossawinery.com

It's fair to say that Spanish exploration in the sixteenth century is the reason that New Mexico was the first state to see *Vitis vinifera* plantings. But it's also fair to add that without Luna Rossa, New Mexico's modern wine industry would not be what it is today. Paolo D'Andrea, a fourth-generation grape grower, was one of a handful of European immigrants who made their way to the Rio Grande river basin in the 1980s. Along with Laurent Gruet, D.H. Lescombes, and Bernd Maier, he is one of a contingent of visionaries who brought their wine knowledge and skills from the Old World to New Mexico. Originally from a small village in the Friuli-Venezia-Giulia region of northeastern Italy, D'Andrea had learned the craft of winemaking and viticulture while at university. He came to New Mexico in 1986 after being hired by a large company in New Mexico to teach the local workforce how to prune the vineyard.

34 THE WINES OF SOUTHWEST U.S.A.

It was here that he met and married his wife, Sylvia, a native of Deming. The two also planted their own vineyard in 1999, and by 2001 they had opened Luna Rossa Winery. D'Andrea's legacy lies not only in Luna Rossa wine but also in his prowess as the state's largest grape grower. In total, he and his team manage more than 330 acres of vineyards (133.5 hectares) in the Mimbres Valley city of Deming, 300 (121.4 hectares) of which is the New Mexico Vineyard, which D'Andrea planted for a private Swiss owner in 1988. The remaining 30 acres (12.1 hectares) form the family's own estate vineyard. As the largest wine grower in the state, it's almost easier to offer a list of New Mexico wine producers who don't buy their grapes from Luna Rossa than it is to share a list of current buyers. They have even sold grapes to a handful of out-of-state producers such as Llano Estacado Winery in Texas and Caduceus Cellars in Arizona. Indeed, without the D'Andrea family's sizable contribution of grapes to the state, the New Mexico wine story would read significantly different today.

Early on, D'Andrea believed Italian varieties were a perfect match for New Mexico's climate and growing conditions and began planting everything from Sangiovese, Barbera, and Dolcetto, to Malvasia Bianca, Refosco, and Montepulciano. Over the years, he has also found promise in Tempranillo, Chenin Blanc, Chardonnay, and Riesling. Though the original intent for the winery was to produce around 3,000 cases annually, production has grown to more than 7,000 cases. All of the Luna Rossa wines are made from the family-grown vineyards and self-distributed throughout New Mexico. In addition to growing grapes, D'Andrea has developed a nursery in which he starts and grows new grapevines to be sold to local area growers as well as for export to Mexico.

Today, D'Andrea focuses primarily on vineyard management and the nursery, allowing his son, Marco, to manage the winemaking. A recent graduate of the University of Udine in Friuli with a degree in enology, Marco has already displayed a deft hand in the winery. Like his father, he finds Italian varieties the best suited to New Mexico's conditions. He has recently employed Ribolla Gialla, a lesser-known white grape from Friuli, as the centerpiece for a Prosecco-style sparkling wine.

The wines of Luna Rossa tend towards full-bodied reds and crisp, vibrant whites evocative of Old-World Italian wines, revealing a common thread among the wines throughout the Southwest. Among Marco's favorites are Aglianico, Negroamaro, and Ribolla Gialla. In the coming years, he looks forward to working with Nero D'Avola, Arneis, and possibly Pecorino. Be sure to try the Reserve Aglianico and Negroamaro, both of which are deeply expressive wines reminiscent of their Italian origins. The Nini red blend of Dolcetto, Nebbiolo, Barbera, Sangiovese, Refosco, Montepulciano, and Aglianico is a consistent crowd favorite.

In addition to the winery and tasting room in Deming, Luna Rossa opened a winery and pizzeria in Las Cruces in 2011. Sylvia D'Andrea invested heavily in the authenticity of the pizzeria's menu, importing a Valoriani Italian brick oven for firing up Neapolitan-style pizzas using fresh local ingredients and house-made mozzarella along with Italian-source San Marzano tomatoes and charcuterie. The restaurant also sells house-made gelato in nearly two dozen flavors and, of course, offers the entire selection of Luna Rossa wines.

MIDDLE RIO GRANDE VALLEY

Gruet
Albuquerque
gruetwinery.com

If there is one New Mexico producer that has stretched its reach and recognition outside of the state borders, it is without question Gruet. For decades, Gruet sparkling wines have been one of the go-to selections for quality American sparkling wine. And while the portfolio of sparkling offerings is indeed well made, Gruet's true legacy lies in the integral part it played in trailblazing a path for today's New Mexico wine industry.

In 1984, Gilbert Gruet of Gruet et Fils wines in Champagne began a search for a place to establish sparkling wine production in the New World. He purchased land in Engle near the Elephant Butte Reservoir and started by planting Chardonnay, Pinot Noir, and Pinot Meunier. As one may expect from a family whose history is deeply rooted in Champagne, Gruet sparkling wines are made according to the traditional method (*méthode champenoise*), which relies on second fermentation and aging in bottle.

Nearly four decades later, Gruet is one of the United States' most nationally recognized sparkling wine producers. The winery works with a handful of vineyard growers in New Mexico but purchases the lion's share of its fruit from California, Washington, and Oregon to bolster its fast-growing market demand. About 80 percent of Gruet's 225,000-case annual production is sparkling wine, placing it well ahead of many other well-known American producers in terms of output. The remainder of Gruet's wine production includes still wines that are sold only within the state. Today, Gilbert's son, Laurent Gruet, is head winemaker for the company, continuing what his father began.

In 2013, Gruet partnered with the Washington-based Precept Wine company but maintained much of its autonomy as a New Mexico wine producer. The overall production today ranges between 150,000 and 175,000 cases of sparkling wine annually. Gruet manages about 75 acres (30.4 hectares) of vineyards, including the Tamaya Vineyard at Santa Ana Pueblo. In addition to sourcing some fruit outside of the state, the producer sources Chardonnay and Pinot Noir from the D'Andreas' Luna Rossa vineyard in Deming. The sparkling portfolio includes brut, *blancs de noirs*, *blancs de blancs*, and demi-sec as well as vintage and limited-edition special releases. And while sparkling wine is the main priority, Gruet also produces limited amounts of still Pinot Noir, Chardonnay, and Pinot Meunier rosé.

A native vineyard: the Tamaya Vineyard of the Santa Ana Pueblo

As New Mexico's wine industry has steadily grown in recent years, it's only fitting that its indigenous heritage is represented as part of its growth. After all, agriculture has been a staple of the culture for hundreds of years. Which is why it felt particularly appropriate in 2013 when the members of the Santa Ana Pueblo, also known as the Tamayame in their own Keres language, decided to plant a 30-acre (12.1-hectare) vineyard on a site a few miles north of Albuquerque. The project came to fruition when the Pueblo's director of agriculture, Joseph Bronk, approached renowned sparkling producer Gruet Winery to partner on the planning and purchasing of the grapes.

Laurent Gruet, head winemaker for the brand, was part of the conversation from the beginning, picking the site for the vineyard at an elevation of 5,110 feet (1,558 meters), one of the highest plantings for Champagne-variety grapes in the world.

The Pueblo of Santa Ana owns and manages the vineyard, making it the first of its kind in the country.[2] All of the grapes harvested from the site ap-

2 While tribes in California and Arizona have purchased existing vineyards, Santa Ana is unique in planting and growing grapes from the ground up.

> propriately named the Tamaya Vineyard, are purchased by Gruet for both sparkling and still wine production.
>
> Gruet wines made with grapes solely from the site, such as the Sparkling Pinot Meunier, are all labeled Tamaya Vineyard and Santa Ana Pueblo, New Mexico. Many of these selections can be found at Gruet's tasting rooms throughout the state, as well as at Prairie Star Restaurant and the Hyatt Regency Tamaya.

Sheehan Winery
Vecinos del Bosque
sheehanwinery.com

Located in Albuquerque's South Valley in the Vecinos del Bosque neighborhood, Sheehan Winery can be found in a converted garage in the backyard of its owner, Sean Sheehan. He began the eponymous label in 2012, using his first harvest from a three-acre vineyard (1.2 hectares) planted in 2009 at Bosque Farms, and officially opened Sheehan Winery in 2015. Sheehan has since added vineyards in Corrales, Bosques Farms, and the South Valley of Albuquerque, and also works with small vineyard growers across the state. With a goal of producing a limited 4,000 cases a year, Sheehan works with everything from Pinot Noir, Tempranillo, and Chardonnay to Cabernet Sauvignon, Cinsault, and the French hybrid grape Chambourcin, selling the majority directly to consumers from his winery.

A graduate of the University of New Mexico with a degree in biology and chemistry, Sheehan originally had aspirations to become a doctor but found himself more drawn to winemaking than walking hospital floors. Production was a slow 10-year process of accruing equipment and tools on a shoestring budget. But in recent years, the winery has hit its stride.

Taking his lead from New Mexico's warm climate and unique soils, Sheehan leans towards making wines that reflect the region's growing conditions. These wines tend to be lighter in style than the wines found in parts of California, where the style is riper, offering both vibrancy and finesse on the palate. Notable selections include the Ollpheist, a unique red blend of heirloom grape varieties Baco Noir and Leon Millot, and the Syrah, with concentrated notes of black and blue fruit, and peppery spice.

OTHER REGIONS

La Chiripada Winery
Dixon
lachiripada.com

Maintaining the name of the original working ranch on which the winery was founded in 1977, La Chiripada is one of the oldest continuously running wineries in New Mexico. Founded by brothers Michael and Patrick Johnson, both originally from California, the winery has built its reputation on a commitment to producing low-intervention wines from New Mexico fruit. Today, winemaking is in the hands of Nathan Johnson, who grew up at the winery and recently took over the reins in the cellar. Located in Dixon, about 50 miles (80.5 kilometers) north of Santa Fe, the winery sits at about 6,100 feet (1,859 meters). It is home to 11 acres (4.45 hectares) of vineyard, primarily planted with cool-climate varieties such as Riesling, Leon Millot, and De Chaunac, due to its high elevation. While a few wines are made from the estate vineyard, the family also sources some of its production from warmer vineyards further south in the Mimbres Valley. The wines are a true reflection of regional terroir, from the racy Kabinett-style Riesling to the beautifully structured Cabernet Sauvignon.

Noisy Water Winery
Ruidoso
www.noisywaterwinery.com

Nestled within the Sierra Blanca mountain range, about 180 miles (290 kilometers) southeast of Albuquerque, the small town of Ruidoso is home to Noisy Water Winery, one of the state's most progressive producers, owned and operated by Jasper Riddle. Though part of a five-generation farming family, Riddle didn't discover his path toward wine until his mid-twenties, following a short-lived career in sports broadcast. At the time, his father, a sommelier, and mother, from an Italian immigrant family who made house wine in their New York barns for many years, were already running their own winery in Ruidoso. Having launched Noisy Water in the early 2000s, their endeavor had started to run out of steam, which gave Riddle a chance to step in and dip his toes into the industry. Following a few courses at the University of California, Davis (UC Davis), Riddle could have easily stuck around to make wine in California. However, his love for his native New Mexico drew him back to begin experimenting with different grapes and winemaking methods. By 2016, when Riddle was just in his early 30s, his wines were turning heads at international wine competitions.

Today, having taken the winery's overall production from 500 cases to more than 40,000, Riddle is considered to be at the forefront of the next generation of New Mexico wine. Employing what he calls a modern approach to classic winemaking, Riddle lightheartedly describes his diverse wine portfolio as a form of madness. He contends that it's because he's not afraid to play around with new varieties and winemaking techniques. You'll find anything from straightforward New World-style Cabernet Sauvignon to orange wines and wild-fermented Pinot Noir in his offerings. He even makes a green chile wine as a local crowd-pleaser.

He sources the majority of his grapes from the Mimbres Valley AVA but farms a few vineyards as well. In 2019 he took over the management of New Mexico's historic Engle Vineyard from Gruet, who had planted the vineyard in the early 1980s. Comprising 250 total acres (101.2 hectares), 125 acres (50.1 hectares) of which are planted with Chardonnay, Pinot Noir, Chenin Blanc, and Cabernet Sauvignon, the vineyard is the oldest continuously farmed vineyard site in the state.

Riddle strives to employ organic farming in the vineyards using ground cover between rows as well as fertilizing with kelp, bat guano, and worm teas.

While his focus has primarily been on traditional wine, Riddle also recognizes the value in catering to the tourist-centric side of New Mexico's wine industry, which attracts novice wine enthusiasts who prefer sweet or exotic wine offerings such as fruit wines and chile-infused wines, a New Mexico hallmark. Riddle knows the sales from these wines help him invest more in the wines he really wants to make. Plus, it gives him a chance to evolve his customers' palates over time.

Though the original Ruidoso tasting room is still up and running, in 2016 Riddle opened a new production facility near Ruidoso in the nearby town of Alto, where he also runs a taproom and event space. Additionally, just a few doors down from the original tasting room, Riddle has opened The Cellar Uncorked as a second tasting room with a different atmosphere. Beyond Ruidoso, you can also find Noisy Water tasting rooms in Cloudcroft, Santa Fe, and Red River.

Vivác
Dixon
vivacwinery.com

Just north of Santa Fe, this stunning winery escape is the creation of brothers Jesse and Chris Padberg, and their wives, Michele and Liliana, respectively. Growing up, the brothers had experimented with making fruit wines on the family farm. Still, it wasn't until shortly after college that the two decided to begin their careers as winegrowers. In 1998, following classes and certifications through UC Davis and the International Wine Guild—an ongoing pursuit, which helps to shape their continued path toward quality wine—the brothers launched Vivác.

The Vivác vineyards lie just beneath the cliffs of the Barrancos Blancos

Located on a family property in the small town of Dixon about 26 miles (42 kilometers) southwest of Taos, the winery is framed by the rugged sandstone cliffs of the Barrancos Blancos, which juts out over the rolling New Mexico scrub-brush terrain. The mountain reveals a picture of the rich mineral content of the region's soils. Spanish for "white canyon" or "white ravines", this mini-mesa, which gradually rises from the southern bank of the Rio Grande, was created when water and sediment became trapped by basalt flows from volcanic uplifts. Compressed by enormous amounts of hydraulic pressure, the towering rock formation emerged when the seas receded millions of years ago, leaving an abundance of alluvial soils behind.

While the majority of their fruit is sourced from Luna Rossa vineyards in the southwestern town of Deming, the Padbergs also farm three separate estate plantings—all of which are organically farmed. In 1999, they planted the Fire Vineyard—which sits at 6,000 feet (1,829 meters) with French hybrids, including Léon Millot, Baco Noir, and Marechal Foch. The 1725 Vineyard sits at 5,800 feet (1,768 meters) and is planted to several varieties, including Grüner Veltliner, Petit Verdot, Pinot Noir, Meunier, Riesling, and red Arrandell. (The vineyard name relates to the year in which Francisco Martin settled the surrounding Embudo Valley.) The third vineyard is the Abbott Vineyard, a four-acre (1.6-hectare) vineyard of Syrah, Merlot, Cabernet Franc, Chardonnay, and Riesling that was planted in 1989.

With the overall philosophy that wine is created in the vineyard, the Padbergs' approach to winemaking is simply to usher the wines along to reveal the best expression of place and vintage. Notable wines include the 1725 Estate Vineyards Petit Verdot with notes of ripe blackberry and baking spices; the Abbott Estate Vineyard Syrah, with smoky tobacco and a silky, blackberry mouthfeel; and the bright and berry-forward Rosé of Sangiovese.

4

WHERE TO EAT AND STAY IN NEW MEXICO

LAS CRUCES

Las Cruces is the largest metropolis in this region, and is one of the fastest-growing cities in the U.S. Considering its appeal for retirees looking for a warm, mild climate to call home as well as for students attending the well-reputed New Mexico State University, Las Cruces serves as a stronghold for winery tasting rooms. Situated beneath the fluted Organ Mountains at the crossroads of Interstate Highways 10 and 25, the sunny town is a hub for travelers from around the country. One of the state's best wine-tasting trails stretches up N.M. Highway 28 from Anthony straight into the center of the city, where visitors can find shops, galleries, and restaurants, particularly in the revitalized historic Downtown area. About 90 minutes northeast of Las Cruces, off of Route 54 are the towns of Alamogordo and Ruidoso, where you can visit another ten New Mexico wineries.

Stay

Hotel Encanto
Las Cruces
hotelencanto.com
Originally built as a Hilton in the mid-1980s, Hotel Encanto is the best in town for an overnight stay. The architecture reflects a classic Mexican hacienda-feel with an earthy color scheme, stuccoed walls and elaborate clay-tiled floors. The quintessential Southwest atmosphere is hard to escape.

Dine

Lescombes Winery & Bistro
Las Cruces
lescombeswinery.com

Luna Rossa Pizzeria
Las Cruces
lunarossawinery.com

Between the Lescombes Winery & Bistro and the Luna Rossa Pizzeria (located almost directly across the street from each other), it's easy to experience both a bit of wine tasting and a satisfying meal. The Bistro offers a varied menu with everything from zesty jalapeño-wrapped shrimp to Asian-inspired potstickers, pastas, and a selection of burgers. At Luna Rossa, authentic Italian is the primary focus. While the wood-fired pizzas are the main draw on the menu, the pasta offerings are also excellent, and the house-made gelato is not to be missed.

See

Mesilla

Use any spare time to wander over to a charming village dating back more than 150 years. Featuring thick-walled adobe buildings housing art galleries and museums, this little town is brimming with tchotchke-filled souvenir shops, but the beautiful historic plaza is worth a look as is the 1855 San Albino Church.

ALBUQUERQUE

At the heart of the Rio Grande Valley is the city of Albuquerque. Nestled against the western slopes of the Sandia Mountains, this historic town was once home to the trade routes forged by the indigenous tribes of the region. Named for the Viceroy of New Spain—the tenth Duke of Albuquerque—in 1706, the city was originally a farming settlement with fertile soils due to its location near the river. One of the largest and fastest-growing towns in New Mexico, this diverse and eclectic historic town offers a range of art galleries, farmers' markets, specialty coffee, craft breweries, and wine tasting rooms, and serves as a perfect gateway to the dramatic New Mexico terrain.

In addition to the pastel-painted Sandia Mountains at sunset, Albuquerque is known for its hot-air balloon festival, its vintage Art Deco buildings, and motel signs along old Route 66; and the pockets of Pueblo heritage peppered throughout the city.

Stay

Hotel Albuquerque at Old Town

hotelabq.com

Reflecting tasteful historic charm, the Hotel Albuquerque at Old Town is an 11-story Southwestern-style hotel situated in the heart of Albuquerque. With New Mexican artisan craftwork on display from Nambe Pueblo-designed metalwork to Navajo rugs, desert-color accents, and hand-wrought furnishings, the Hotel Albuquerque puts guests in a distinctive Southwest state of mind.

Hotel Chaco

hotelchaco.com

Located in the urban Sawmill district of Albuquerque's Old Town, Hotel Chaco is a soulful homage to the historic Chaco Canyon, designed to evoke the fine stone chinking that comprises most of the ninth- to twelfth-century structures found at the treasured ancient Puebloan site. Rooms are spacious and artfully decorated with a nod to the region's indigenous heritage blended with a distinctive modern flair.

Dine

Crafted at Hotel Chaco

hotelchaco.com

Featuring a curated selection of New Mexico wines and spirits from producers such as Gruet, Noisy Water, Vivác, and Santa Fe Spirits, Crafted is a recent addition to the Sawmill district, located in the Hotel Chaco. The tasting room's menu features wines by the bottle and glass, as well as hand-picked flights that spotlight the unique and native flavors in each category.

Garduños

gardunosrestaurants.com

A festive respite located in the Hotel Albuquerque, Garduños offers a festive yet casual atmosphere with a Mexican-inspired menu that gives an authentic taste of Southwest flavor. Enjoy nachos, street tacos, chile rellenos, or guacamole made table-side, and sip one of the specialty margaritas from the bar.

Gruet Bubble Bar at Level 5

hotelchaco.com

This rooftop bar of the Hotel Chaco celebrates the allure of fizz, featuring a selection of sparkling wines from the Gruet portfolio. Guests can sip, and stroll the rooftop patio while taking in the best views of the city.

Level 5

hotelchaco.com

Located on the fifth floor and rooftop of the Hotel Chaco, Level 5 is named for Chaco Canyon's Pueblo Bonito, a preserved ancient Pueblo, which has five levels and serves as one of the most treasured remnants of Puebloan culture. French chef Christian Monchâtre expertly executes a balance between classic techniques and indigenous regional and Latin American fare using local ingredients.

SANTA FE

Perhaps the most famous destination in the state, Santa Fe is a Mecca of the American Southwest, paying homage to the region's many cultural influences over the past several centuries. Situated on a 7,000-foot plateau at the base of the Sangre de Cristo Mountains, the charming town boasts an endless selection of art galleries, excellent restaurants, a plethora of Southwestern furniture and jewelry shops, and beautiful adobe architecture throughout. The town's central Plaza hosts numerous events throughout the year and has been the site of bullfights, gunfights, political rallies, promenades, and public markets since the early seventeenth century.

Founded in the early 1600s as "La Villa Real de la Santa Fe de San Francisco de Asísi" (the Royal City of the Holy Faith of St. Francis of Assisi), Santa Fe was claimed for Spain by Don Pedro de Peralta. Despite a storied past of both struggle and peace between Spanish settlers and indigenous peoples, the city has evolved into a twenty-first century bastion of art and culture. It was the first American city to be designated a UNESCO Creative City, acknowledging its place in the global community as a leader in art, crafts, design, and lifestyle.

Route 66: a highway of years gone by

It has been called the path to Western promise by Oklahomans escaping the Dust Bowl in the 1930s, the "mother road", and the "road of flight" by classic American novelist John Steinbeck in his Depression-era tale, *The Grapes of Wrath*. It served as the inspiration for the Beat Generation of writers and was extolled by Jack Kerouac in his classic, *On the Road*. But it was the catchy 1946 tune, "Route 66", written by Bobby Troup and recorded by the likes of Nat King Cole, the Rolling Stones, and Depeche Mode that imprinted the fabled roadway in the minds of many throughout the world.

The historic U.S. Route 66 was a journey that ran east–west through eight states from Chicago, Illinois, to Los Angeles, California, including the central part of New Mexico and Arizona. Long before it had earned its beatnik cool factor, the corridor was one of the first transcontinental highways in the country—even before it was paved in 1926. During the 1930s and until more than a decade after the Second World War, Route 66 earned the title "Main Street of America" for its close connection to the small-town culture of the Midwest and Southwest, lined as it was with hundreds of cafés, motels, gas stations, and tourist attractions.

The Dust Bowl was a time of devastating drought in the early 1930s that saw a wave of travelers heading through the Southern Plain states for the West Coast. Another great migration took place after the Second World War, when thousands of Americans left the confines of the industrial East for the idyllic coastal appeal of Southern California, marking a significant demographic shift from the Rust Belt to the Sun Belt.

Passage along Route 66 experienced a steady decline after 1956, when President Eisenhower introduced the Federal-Aid Highway Act, inspiring the construction of large, interstate highways, much like the German *autobahn*. By the 1970s, the route was largely replaced by five different interstates, including Interstate 40, which serves most of the Southwest. By 1985, the entire route had been decommissioned and was later designated as Historic Route 66.

Though it is no longer a main route across the country, Route 66 has retained its mystique, with displays of neon signs, rusty middle-of-nowhere truck stops, and kitschy Americana. Tourists looking to graffiti half-buried Cadillacs can still visit Cadillac Ranch in Amarillo, Texas, or take an easy ride like Peter Fonda and Dennis Hopper did on their motorcycles. As the song suggests:

Won't you get hip to this timely tip
When you make that California trip
Get your kicks on Route 66.

Stay

El Rey Court

elreycourt.com

For a hip, low-key experience on a slice of historic Route 66 just a few miles outside of town, El Rey Court is the place. This refurbished 1936 motor court has a mid-century modern slant, and the Southwest chic feel of the lobby is enough to keep you lingering over a house-made margarita from the adjacent La Reina bar for hours. Be sure to grab a sensational diner-style breakfast at the nearby Pantry Restaurant.

Four Seasons Rancho Encantado

fourseasons.com/santafe

Offering the standard of luxury you'd expect from the Four Seasons brand, Rancho Encantado also imparts an authentic Southwestern feel best described by the meaning of its name, "The Enchanted Ranch". Set across 57 acres (23 hectares) at the foothills of the Sangre de Cristo mountains, just on the outskirts of Santa Fe, the resort features 65 Casita guest rooms and suites, a full-service spa, and an array of cultural and outdoor experiences, including visits to the ancestral lands of indigenous peoples, mountain biking, hiking treks, fly fishing, river rafting, snowshoeing, and skiing. The resort spa offers a menu of wellness options that blend native healing practices and ingredients with modern techniques.

Hotel St. Francis

hotelstfrancis.com

Located in the heart of town, the historic Hotel St. Francis is a charming boutique hotel named for St. Francis of Assisi. With its sleek whitewashed adobe walls and Mission-Revival décor, accommodations are Old-World luxury with a contemporary twist. In partnership with the hotel, Gruet has a tasting room adjacent to the hotel lobby. The atmosphere is laid-back, yet intimate, and guests can stop in for a full tasting of Gruet's sparkling as well as still wines, or simply enjoy a glass alongside a cheeseboard with friends, at the bar or on the outside patio.

Dine

Coyote Cafe

coyotecafe.com

A stalwart in the world of Southwestern cuisine, Coyote Cafe is known for some of the most extravagant and delicious cuisine in the city. Items such as porcini mushroom soup with green chile albondigas (meatballs) and Tacos Al Pastor are some of the menu classics. For lighter bites in a more relaxed atmosphere, the adjacent Coyote Cantina, a fun rooftop space, is a worthwhile alternative. Here, you can expect burgers, pork carnitas tacos with pineapple-habanero salsa, and craft cocktails.

Luminaria

hotelloretto.com

The pretty patio at Luminaria looks out at the Romanesque architecture of the Cathedral Basilica of Saint Francis of Assisi, or Saint Francis Cathedral. Casual, yet sophisticated, Luminaria offers a Southwest-inspired menu with specialties such as grilled achiote-marinated shrimp with pumpkin mole and wood-fired steaks.

Sazón

sazonsantafe.com

The restaurant offers an upscale take on regional Mexican fare, but also worth a look is the extensive list of artisan tequilas and mezcals. The mole is not to be missed; each day Chef Fernando Olea features a different mole served with local meats or fresh fish. The chef's degustación tasting menu, with seatings at 5:30 pm and 8:00 pm, is the best way to experience the depth of flavors achieved by Olea's kitchen.

The Shed

sfshed.com

Since 1953, The Shed has been quite possibly the hottest ticket in town during the lunch hour. Serving some of the most flavorful New Mexican food and margaritas around and housed in a seventeenth-century adobe building adorned with folk art, The Shed offers warm and friendly service. Crowd favorites include red-chile enchiladas, green-chile stew with potatoes and pork, and the charbroiled Shedburgers.

Terra

fourseasons.com/santafe

Just outside Santa Fe, Terra at the Four Seasons Rancho Encantado offers a serene dining experience featuring a chef-driven menu using local and garden-grown ingredients. Dishes include spicy green chile corn chowder, rack of lamb with blue corn polenta, and fire-grilled ribeye steak, and guests can take advantage of hands-on cooking class options as well. An excellent spot for a getaway and an exceptional dining experience.

See

Galleries

With nearly 200 art galleries sprinkled throughout the city, it's almost impossible to take all of them in during one visit. The regular First Friday Night Art Walks are a great way to experience the assortment of ceramics, paintings, photography, and sculptures. Be sure to stop in at the Georgia O'Keeffe museum, featuring the work of this prominent American Modernist artist, who drew most of her inspiration from the landscape of the area.

Kasha-Katuwe Tent Rocks National Monument

Jemez Springs
blm.gov/visit/kktr

For a unique outdoor excursion, drive to this National Monument. About an hour west of Santa Fe, it features unusual sandstone rock formations resembling stacked tents in a stark, water- and wind-eroded box canyon. Hiking is an option year-round, although the summer season can be sweltering.

Meow Wolf

meowwolf.com

Meow Wolf's House of Eternal Return is an explorable, immersive art installation that's impossible to explain in words. Visitors are left to wander through a warehouse maze of whimsical art creations; it's like taking a journey through the pages of *Alice's Adventures in Wonderland*. Reserve tickets online in advance to secure a visiting slot—the earlier in the day, the better.

The Turquoise Trail

Peppered with ghost towns, piñon pine forest, and sweeping mountain vistas, the Turquoise Trail is a worthwhile alternative route to take from Santa Fe to Albuquerque, avoiding the busy interstate highway. Since the 1970s, this nearly 70-mile (112.5 kilometers) stretch along the N.M. 14 State Highway on the eastern side of the Sandia Mountains may set you at a more leisurely pace, but as you wind your way along the rolling foothills, you'll not only eperience panoramic views of the Ortiz, Jémez, and Sangre de Cristo Mountains but also discover one of New Mexico's most treasured trade commodities: turquoise.

The brilliant bluish-green opaque stone is a rare mineral that has been the centerpiece for art, jewelry, and the overall color scheme for much of the region. It has long been found throughout parts of Iran, the Sinai Peninsula, and China, but the American Southwest is also a significant source, particularly Arizona, Nevada, and New Mexico. Evidence suggests that indigenous peoples mined turquoise in this particular area more than 2,000 years ago; they were followed by the Spanish, who began mining it in the sixteenth century. The hunt for the semi-precious stone continued into the early twentieth century with individual artisans and large retailers, including Tiffany & Co., seeking to profit from the brilliant blue stone.

Its historic appeal has distinguished the route as an official National Scenic Byway, and it has attracted many writers, artists, and other urban refugees over the years. You can drive the loop in three hours with minimal stops, or make a full day of it, taking in the views from the top of Sandia Crest and stopping off in the mining towns of Golden, Madrid, and Cerrillos.

PART 2

TEXAS—WINE IN THE LONE STAR STATE

5

HISTORY

Before European exploration, Texas was essentially the crossroads of two major North American cultural spheres, the Southwestern and Plains areas. During pre-Columbian times, the region was occupied by three major indigenous cultures: the Pueblo in the west, the Mound Builders of the Mississippi culture to the east, and the Teotihuacan to the south. These cultures would evolve into six groups by the seventeenth century, including the Caddoan peoples who occupied the area around the Red River in the northeast, the Atakapa tribes along the Gulf Coast, the Puebloan people along the Rio Grande and the Pecos River in the west, the Apachean tribes in the northwest, and the Comanches in the northern part of what is now the Texas Panhandle.

Texas' recorded history began with Spanish exploration along the Gulf Coast by Alonso Álvarez de Pineda, a cartographer who led an expedition throughout the Gulf of Mexico, from Florida to Texas, mapping the coastline along the way. Although Pineda would claim the area for Spain, it would be more than 150 years before Texas would be explored again.

In 1684, under the leadership of nobleman René-Robert Cavelier, Sieur de La Salle, a French exploration made its way west from Louisiana. Although his original mission was to sail from France to claim the entire Mississippi river for his country, navigational miscalculations led La Salle's party to the shores of Matagorda Bay, where they built Fort Saint Louis. La Salle's foray onto Texas soil was not long-lived. Within a decade of his arrival, the settlement had fallen to native tribes; La Salle himself perished in a battle in east Texas. Though the Spanish also attempted to establish missions in east Texas in the late seventeenth century, all attempts failed. Over the next 110 years, Spain established numerous villages and missions in the province, while continuing to meet resistance from native tribes.

During the early nineteenth century, Texas was largely considered a buffer zone between Spain and the expanding United States, but in 1819 Spain gave Florida to the U.S. in return for undisputed control of Texas. Mexico then led an uprising against Spanish rule, coopting Texas as part of its territory. In 1821, it won its independence, making Texas a part of the new Mexican nation. To encourage settlement, Mexican authorities allowed organized immigration from the United States, and by 1834 more than 30,000 Anglos had settled in Texas. The newly established Mexican Texas would last until 1836 when Texans launched their own revolution, establishing independence from Mexico as the Republic of Texas.

However, the newly formed republic was unable to defend itself from further incursions by Mexican troops and eventually negotiated with the U.S. to join the Union in 1845, becoming the twenty-eighth state in the Union. Texas declared its secession from the United States in 1861 to enter the Confederate States of America in the Civil War, rejoining the Union in 1865 at the war's conclusion.

Until the mid-nineteenth century the Comanche reigned as the dominant tribe of the Southwest, particularly in the central and northwest quadrant of the state (and beyond to New Mexico and northern Mexico). The Comanche fended off significant occupation and settlement in the region. However, contact with this new culture led to epidemics of smallpox and cholera within the indigenous communities. The epidemics along with military campaigns by both the U.S. army and Texan forces led to the eventual demise of the native empire.

TEXAS WINE HISTORY

While New Mexico claims the distinction of being the first state to have *Vitis vinifera* planted, in 1629, it's fair to say that Texas was likely the second. Those first vines were planted within miles of the border of the two neighboring states during early Spanish mission settlements, and it's worth noting that the territory was all part of Mexico at the time, and had technically been claimed by Spain, so the distinction is fairly new. The first settlement around El Paso was established in 1659, and according to Thomas Pinney in his book *A History of Wine in America: From the Beginnings to Prohibition*, by 1744 one Franciscan monk was writing that orchards and vineyards were flourishing.

In the early nineteenth century, wines from El Paso, which are referred to as "Pass Wine", became popular among Santa Fe traders. In fact, according to Pinney's book, El Paso wine had become the only revenue-producing crop in the whole province of Spanish New Mexico, of which El Paso was a part at the time. Wines during this era were in one of three styles: a high-quality light red wine; a lower-quality darker wine, the Pass Wine; and a lesser-quality, low-price wine.

East of El Paso, much of the state had already begun experimenting with native varieties such as the Mustang grape (*Vitis candicans/Vitis mustangensis*), which was said to make tolerable red wine and decent Port-style wine. In addition to Mustang, Muscadine (*Vitis rotundifolia*) and Lenoir (also known as Jacquez or Black Spanish) saw a rise in popularity. Although no significant commercial production of wine ever came to pass, the use of wild and hybrid varieties continued among local hobbyists throughout much of the nineteenth century. *Vitis vinifera* plantings were virtually non-existent by the end of the century except for a 200-acre (80.9-hectare) planting by Italian immigrants near the southern Mexican border town of Laredo.

Independent of this Italian planting, another Italian family also began operation in South Texas about 200 miles (322 kilometers) north of Laredo, in Del Rio. Here, Italian immigrant Frank Qualia took note of the native Lenoir grapes growing wild in the region and used cuttings to plant a family vineyard in the calcium-rich loamy soils of South Texas. In 1883, he opened Val Verde Winery, named for the county in which it is located. Today, the fourth-generation operation can claim the distinction of being the oldest continuously operating winery in Texas, having survived Prohibition by making sacramental wine for church services.

While Qualia's vineyards were taking root, another key Texas wine figure was making his own impact more than 500 miles (805 kilometers) away in the far northeast side of the state. Just northeast of Dallas in Denison, Texas, Thomas Volney Munson, a grape horticulturist, spent nearly four decades studying, collecting, and hybridizing native American grapes. Having first studied at the University of Kentucky and familiarized himself with the native varieties of the Midwest, Munson later spent time working with native varieties in the frigid

58   THE WINES OF SOUTHWEST U.S.A.

climes of northwest Nebraska. But it was Texas that he ultimately made his home, and his work would define him as the foremost expert in America's native grapes. Later, in the late nineteenth century, he would go on to provide the specific rootstocks that were resistant to phylloxera (see box, p. 60), and ultimately save the whole of the European wine industry, an achievement for which he would receive the Chevaliers du Mérite Agricole in the French Legion of Honor in 1888. Today, nearly all grapevine rootstocks used worldwide rely heavily on native Texas grape species.

During the early part of the twentieth century, grape growing and winemaking never really took off. It didn't help that Texas passed state-wide Prohibition in 1919, a full year before the Eighteenth Amendment was put into effect nationwide. Add to this the fact that when the national law was repealed with the Twenty-First Amendment in 1933, it didn't apply in Texas due to the precedence of state law. In 1935, the Texas Legislature, in a special session, passed the Texas Liquor Control Act, which provided that local jurisdictions reverted to their pre-Prohibition, local-option status, whether wet (alcohol sales allowed) or dry (alcohol sales not allowed). This left most Texas counties, and many smaller juris-dictions, entirely dry. Bit by bit, counties and municipalities would begin to establish laws allowing for the sale of liquor, wine, and beer. A quick look at the Texas Alcoholic Beverage Commission website shows how dry counties have dwindled in recent years.[3]

Indeed, Prohibition caused a significant setback to the wine industry. Immediately following its repeal, larger wineries in California ramped up production to flood the market with a glut of wines that valued quantity over quality. In the meantime, individual states were given complete control over their own alcohol laws, with many opting to re-main dry for much longer. In the Southwest states in particular, there was a dearth of wine production for decades leading up to the 1960s.

Texas viticulture did have pockets of activity. From 1938 to 1995, C.O. Forester, Jr., of Elsa, Texas, hybridized grapes in search of a ta-ble grape for the Lower Rio Grande Valley. Researcher Norman Wilms tracked the success of two hybrid varieties in the Lower Rio Grande Valley from 1966 to 1995. The first variety was Muscanal, a natural hy-brid of *Vitis candicans*, which demonstrated resistance to fruit and vine diseases. The second was Convent, appropriately named for the two

---

3   In 1995, there were 53 dry counties. In 2003, there were 35 completely wet counties and 51 completely dry. As of 2018, there were only 5 completely dry counties.

French nuns who brought it with them from France. It still grows today at the Immaculate Conception Cathedral in Brownsville, Texas. In the 1950s, Mr. J.H. Dunn planted an experimental vineyard in Lubbock in the Texas High Plains. Around the same time in south Texas, a large number of commercial table grape vineyards were planted from Crystal City south to Rio Grande City, totaling more than 3,000 acres (1,214 hectares). Unfortunately, all of these vineyards succumbed to Pierce's disease, cotton root rot, and possibly freeze injury.

The story of the modern Texas wine industry began in the late 1960s in both the north and central parts of the state. Several key players helped contribute to its genesis. In 1973, Texas A&M University offered George Ray McEachern, an extension horticulturist, a grant to place trial vineyards across the state using 12 grape varieties. He selected a total of 30 different test plots in Fort Stockton, San Antonio, Abilene, and Texarkana, among others. For each location, he planted three American, three hybrids, three *Vitis vinifera*, and three varieties of the grower's choice.

Around the same time, researcher Ron Perry began performing soil and climate studies for A&M, and Texas Tech chemistry professor Dr. Roy Mitchell started a research program on winemaking. In many ways, this is where Texas wine's first chapter really begins, in the High Plains region near Lubbock. The area is more commonly associated with tumbleweeds, oil derricks, and row crops; vineyards didn't come into view until the late 1960s, when a couple of researchers, Robert Reed and Clinton "Doc" McPherson at Texas Tech University, planted a few vines. (McPherson planted Sangiovese, which still grows today under the management of his son, Kim McPherson.)

In the following years, other Texas wine pioneers jumped into the game, inspired by the national acclaim California was receiving in the late 1970s after its hard-fought efforts to be taken seriously by the global wine community. McPherson went on to become one of the founders of Llano Estacado Winery in Lubbock, which remains one of the oldest and largest wineries in the state. Further south, in the Hill Country of Central Texas, Ed and Susan Auler of Fall Creek Vineyards planted Bordeaux varieties with the encouragement and tutelage of famed winemaker André Tchelistcheff in the late 1970s. Around that same time in Bryan, near Texas A&M University, the young Paul and Merrill Bonarrigo were eager to discover how vineyards would fare in the south-central part of the state.

How Texas helped conquer phylloxera

In the late nineteenth century, the spread of phylloxera through Europe was slow and ruinous. The solution proved to be grafting European vines onto American rootstock. It's an interesting byway of Texan wine history that it was a Texan who helped in those efforts.

The first wave of French experimentation with grafting to American rootstock failed due to the selected rootstocks being more suitable for fertile soils. But by 1880, phylloxera had overtaken nearly 2 million acres (more than 800,000 hectares) of French vines. In 1887, a contingent of French researchers convinced the French government to approve a second attempt at American viticulture. This initiative was launched under the direction of Professor M. Pierre Viala from the Institut National Agronomique in Montpellier. The fact-finding mission was more specific in nature, focusing solely on finding an American rootstock variety that could grow in similar soils to those of southern France. Viala traveled to Washington, D.C., but his initial few months in America failed to reveal the best varieties. Much of his research included areas of Pennsylvania, Virginia, and Tennessee, all of which lacked the same mineral composition of French soils. As his survey brought him further west, he would discover that the Lower Cretaceous limestone formations of central Texas held promise.

Viala reached out to reputed grape horticulturist Thomas Volney Munson (see p. 57) in Denison, in the northeast part of Texas: this vast landscape was his particular area of expertise. Viala and Munson had already corresponded for some time on the subject of phylloxera, and the French professor was eager to spend several weeks in Texas. Munson shared his extensive research on native Texas grape varieties with Viala in Denison before sending the expedition further south to the region where they were likely to find their most encouraging discoveries, the Hill Country of central Texas. Here, in the poor, calcium-rich soils of the area, Viala found the variety Munson believed would provide the ideal solution, *Vitis berlandieri*. At Munson's suggestion, he also studied the *Vitis cinerea* and *Vitis cordifolia*, variations of which he had already encountered in Tennessee and Missouri. Still, he was excited to see them thriving in the limestone-rich soils of the Texas Hill Country. In his final report, he recommended these three Texas-specific species as the grafts most likely to thrive in France, noting that "this [saving the French vines from phylloxera] can only be done by grafting upon native Texas vines."

The next task was to provide an ample supply of these varieties to southern France. Munson organized dozens of workers and landowners to collect 15 wagons of dormant stem cuttings, in addition to tens of thousands of cuttings from his own vineyards for shipment to France. The results proved promising. *Vitis berlandieri* proved particularly valuable, though not without faults. Many producers had difficulty rooting the cuttings, meaning not all of the vines survived when planted. Still, work with the native species was not all lost. The French used the *Vitis berlandieri* and other Munson-recommended grapes as well as American natives *Vitis rupestris* and *Vitis riparia* to create hybrid rootstocks for grafting, a solution which proved successful. Today, the majority of grapevines in France, as well as throughout Europe, are grown on these resulting grafted rootstocks. In the case of *Vitis berlandieri*, specifically, the rootstock crossed with Chasselas resulted in the highly regarded rootstock 41B, which would replant both the Champagne and Cognac regions. Other crosses included *Vitis berlandieri* with Cabernet Sauvignon to yield the Number 333 School of Montpellier and the Number 99 Richtor, 110 Richter, 1103 Paulsen, and 140 Ruggeri crossings. In the coming years, other American stocks would also provide manageable options.

Munson's research and contribution to solving the French phylloxera dilemma would earn him the Chevaliers du Mérite Agricole in the French Legion of Honor in 1888, a distinction which is typically reserved for French Nationals. He later received national recognition for his comprehensive Exhibit of American Grapes at the 1893 Exposition in Chicago, Illinois. The Exhibit was subsequently given to the United States Department of Agriculture in Washington, D.C., and was the most extensive and most accurate single collection of grape species ever made. By 1900, it is estimated that two-thirds of French vines were bound to American roots. Today, nearly all grapevine rootstocks used worldwide rely heavily on native Texas grape species.

Munson not only classified grapes, but also collected a vast number of native vines and other varieties, which he bred and evaluated for selecting outstanding cultivars. This was one of the most exceptional private plant breeding programs ever developed in the country. Munson's research continues at Grayson College in the Sherman-Denison area of northeast Texas. In 1975, the college established the Thomas Volney Munson Memorial Vineyard in recognition of his contributions to horticulture. The vineyard continues to cultivate and preserve some of the varieties Munson discovered. In 1988, the college launched the T. V. Munson Viticulture and Enology Center next to the vineyard.

62 THE WINES OF SOUTHWEST U.S.A.

Throughout the 1980s and early 1990s, the infant industry slowly attracted other participants to the game, including Pheasant Ridge and CapRock in the High Plains as well as many in the Hill Country, which was enjoying an influx of tourism to the quaint German-heritage towns of Fredericksburg, Kerrville, Comfort, and New Braunfels. Producers such as Becker Vineyards, Grape Creek Vineyards, Fredericksburg Winery, Sandstone Cellars, and Flat Creek Estate, all contributed to a growing interest in answering the question of whether or not wine production could find a footing in Texas. But while there have been a few blips of success in the past, with national awards and a growing interest among Texan wine consumers, the defining era of Texas wine is arguably right now. More than a decade of focus and refinement from progressive winegrowers and producers has elevated Texas wine into a promising new era.

Today, Texas is home to about 500 wineries, according to the Texas Wine and Grape Growers Association. It's important to note that this number may be misleading since it is based on the number of winery permits issued by the Texas Alcoholic Beverage Commission as of 2019. Not all of these permitted businesses are producing bottles or functioning as producers of actual Texas wine. For instance, some Texas restaurants file for a winery permit to be able to sell wines under their own label that aren't necessarily Texas wines. Other wineries merely buy bulk wine from other states such as California or Washington and sell it out of a tasting room. These businesses are still required to have a winery permit, but they do not contribute to the actual production of Texas wine. In addition, some existing Texas producers have opted to open up second, third, or fourth locations, each of which requires a separate winery permit. These numbers make it difficult to get a true picture of who is making Texas wine. The total number of single businesses producing wines made from actual Texas grapes does not have an official reporting entity but is estimated to be around 350 total wineries. But with an estimated 6,000 acres (2,428 hectares) under vine and more than 14,000 tons (12,701 metric tons) of grapes produced each year, it is clear that in the realm of wine industries across the country, Texas has become a contender.

German heritage in the Texas Hill Country

In the mid-nineteenth century, central Texas became the scene for significant German immigration as part of the Mainzer Adelsverein, or Nobility Society of Mainz, which aimed to establish a new Germany on Texan soil in 1842. New Braunfels was the first town built under the movement. In 1845 Prince Carl of Solms-Braunfels led hundreds of sea-lagged German settlers from the Gulf Coast city of Galveston to a plot of land north of San Antonio on the banks of the Comal River. The settlement, named New Braunfels for his hometown in Germany, Braunfels, endured a shaky beginning with the outbreak of the Mexican–American War in 1846. However, by 1850, the town was a thriving community, boasting the title of the fourth largest city in Texas.

Fredericksburg, which is named for Prince Friedrich of Prussia, was the second of the Mainzer Adelsverein establishments, founded by Baron Ottfried Hans von Meusebach, also known as John O. Meusebach, in 1846. Meusebach became famous for brokering a peace treaty with the Comanches, which prevented raids on the town. The treaty also encouraged trade in the area, and to this day, it is the only American–Indian treaty in the state that has not been broken. Cattle and agriculture eventually became the primary sustainable commerce in the city as it grew through the Civil War and continued into the twentieth century. Today it is the epicenter of the Hill Country wine scene.

In 1852, the town of Comfort was founded by Ernst Hermann Altgelt. It was here that a community of German independents known as the "Freethinkers" settled, in search of refuge from political and religious oppression. The town remained a peaceful place until the outbreak of the Civil War. While most Texans joined the Confederate Army in resistance to the United States, the Freethinkers sided with the Union army. Fearful of threats from Confederate loyalists, much of the community fled toward the Mexican border for protection. But 36 Freethinkers met their doom at the 1862 Nueces Massacre, a violent battle between Confederate soldiers and German Texans. Today Comfort is the home of one of only six flags across the country that flies at half-mast year-round in remembrance of those massacred.

6

CLIMATE, REGIONS, AND CHALLENGES

CLIMATE

Considering its size, it should come as no surprise that Texas is home to a range of climate types ranging from arid and semi-arid in the west to humid and subtropical in the east. Various weather patterns develop depending on location.

In the northern plains, it is semi-arid with average annual precipitation between 18 and 20 inches (457–508 millimeters). The northern Panhandle, so named for its resemblance to the handle of a large frying pan, is one of the coldest areas in Texas and receives the most snow in the state (about 25–30 inches annually, or 635–762 millimeters). While the winters are cold and dry, the summer season is hot and usually doubly dry.

The rolling hills of central Texas are marked by a semi-arid climate in the western part of the region, and more of a subtropical climate in the rest. Summers are hot and humid, while winters are mild. The west side of central Texas receives an average annual rainfall of 21 inches (533 millimeters), while the rest of the region gets a significantly higher amount with an average of 35 inches (889 millimeters).

Far west Texas borders New Mexico and exhibits similar climate conditions, with an arid, desert-like environment. The annual precipitation here is the lowest in the state with approximately 15 inches (381 millimeters), though the mountain ranges can experience heavy snowfall. Summers are dry, while the rest of the year in this region has the clearest skies in Texas.

Though fewer grapes are grown in east Texas, this region does have a reputation for producing quality hybrid varieties. The climate here is significantly more humid and subtropical, receiving an annual rainfall of about 60 inches (1,524 millimeters), the highest amount in the entire state. Severe thunderstorms, hurricanes, and tornadoes are frequent in the region overall during the spring season; the summer is hot and humid in the east, while the coastal sections to the south receive cooling breezes from the Gulf of Mexico.

As the state is accurately associated with sweltering heat, it's not surprising that the challenge during the growing season is a condensed time frame. Depending on where you are in the state, bud break may begin in early March, and in recent years, harvest has started as early as the last week of July.

> ## Caliche soils in the Southwest
>
> When it comes to similar soil types in the Southwest, one common thread you'll find across this arid region is the presence of caliche, a whitish-gray or cream-colored soil layer of cement material, typically made up of calcium carbonate. The term "caliche" is Spanish, originating from the Latin "calx", meaning lime. Caliche is known by many other names, the more common of which are calcrete, hard-pan, duricrust, and calcic soil.
>
> This unique material is formed when carbonates in the soil are dissolved and leached by rainwater. In arid regions such as the Southwest, the water rapidly evaporates, or is absorbed by vegetation, leaving behind the carbonates to bind with other soil particles such as gravel, sand, clay, or silt. The result is either a soft, thin soil layer, often coating rocky soils, or a solid, hard-pan layer, ranging from a few inches to several feet in thickness. The softer caliche can impose some nutrient-related issues for crop growth, due to its elevated pH and the presence of bicarbonates, while the hard-pan layer offers poor soil drainage and poses a difficulty for root penetration. (Note that though the cement is primarily made of calcium carbonate, other known types of cement include magnesium carbonate, gypsum, silica, iron oxide, or a combination of these materials.)
>
> Caliche layers are often found in calcareous soils with high pH, ranging from 7.5–8.5. These alkaline soils can limit the availability of phosphorus, iron, boron, zinc, and manganese since the solubility of these nutrients is reduced at high pH. This makes it necessary for grape growers to monitor micronutrient levels in the vines and amend nutrients as needed, particularly during the growing season.
>
> Occurring commonly throughout the Southwest, caliche is also found in central and western Australia, in the high desert terrain of the Andes mountains in South America, in parts of Central America, including the Sonoran and Mojave deserts, and in eastern Saudi Arabia.

REGIONS

The phrase "everything is bigger in Texas" isn't an exaggeration, especially when you consider that the state is roughly the size of France and Switzerland combined, plus a little bit of Germany. At 268,820 square miles (696,240 square kilometers), Texas surpasses France in area by more than 20,000 square miles (51,800 square kilometers). Knowing this makes it easier to put Texas in perspective when defining its wines

68 THE WINES OF SOUTHWEST U.S.A.

and wine regions. Just as France sees a diversity in climate and soils, from Champagne to Bordeaux, so does Texas.

In general, the geologic history of the state is characterized by uplifts of mountains, inundation by vast inland seas, river transports of large volumes of eroded sediment, volcanic eruption, and earthquake activity. All of these occurrences have contributed significantly to the geologic development of Texas. Petroleum resources, lignite, metal ores, groundwater, salt, limestone, granite, ceramic clays, and fertile soils are all part of the legacy of the changing face of Texas.

When it comes to Texas wine, there are officially eight American Viticultural Regions (AVAs), but there are two regions that claim the lion's share of attention, and for very different reasons: Texas Hill Country AVA, and Texas High Plains AVA.

Nowhere but in Texas—the physiographic regions of the Lone Star State

The saying "everything is bigger in Texas" may be a tired old cliché, but in some cases, it's a fact that cannot be overstated. Take, for instance, the state's geography. Of the eight major physiographic regions in the United States, five of them are found in Texas. When it comes to climatic variation, wildlife diversification, and vegetational striation the state is, without question, literally at the crossroads of the country, where east meets west, and north meets south.

The Gulf Coastal Plain spreads out over the southern and eastern portions of Texas. Formerly an expansive ocean floor, the region extends north into the state all the way to the Red River (around the Dallas–Fort Worth area and on into Oklahoma) and to its eastern borders. It is bound on the west by the Rio Grande and along the Balcones Escarpment, which defines central Texas. Elevations within the Gulf Coastal Plain range from around 950 feet (290 meters) to sea level.

The Interior or Central Lowlands is a continental physiological region that stretches from Canada to west-central Texas. This region is essentially the area west of Dallas, including Fort Worth, which stretches west to the Caprock Escarpment, which begins the Texas High Plains. It is bordered to the south by the Llano Basin and the Edwards Plateau, which define the Texas Hill Country. Within this region, elevations range from 1,000 feet (304 meters) at its eastern edge, to 2,200 feet (670 meters) in the west, at the base of the Caprock Escarpment.

The High Plains and Plateaus region extends throughout west Texas and is primarily defined by the Llano Estacado. While the High Plains are typically thought of by Texans as being in the northern part of the state, the actual physiographic region reaches further south into central Texas. It may surprise some that the Texas Hill Country is technically part of the formal High Plains and Plateaus region. While the Llano Estacado defines the northern part of the High Plains, the southern region is determined by the Llano Basin, the Edwards Plateau, the Toyah Basin, and the Stockton Plateau. Elevations throughout the northern High Plains range between 2,500 and 4,500 feet (762–1,372 meters). In the southern part of the High Plains, which is separated from the Llano Estacado, elevations range between 850 feet and 4,000 feet (259–1,219 meters) at the foot of the mountains west of the Pecos River.

The Basin and Range Province, which is also referred to as the Great Basin, lies in the far west stretches of the state and is reminiscent of the terrain most associated with the American West. Arid, high desert flats are met by mountain chains, including the Davis, Van Horn, Franklin, Glass, and Guadalupe Mountains.

The southernmost tip of the Rocky Mountains region extends into the very western corner of the Texas–New Mexico border. Guadalupe Peak serves as the highest elevation in Texas at 8,749 feet (2,667 meters).

Notable geographical formations within the Texas wine regions
Llano Estacado. The key physiographic feature of the Texas High Plains AVA, this elevated flat plain bears a Spanish name, which means "staked plains". The formation lies at the southern tip of the High Plains of North America and serves as one of the largest tablelands on the continent, canvasing 32,000 square miles (82,880 square kilometers); the formation is larger than all of the states of New England combined. This high mesa slopes at a rate of approximately ten feet per mile (3 meters per 1.6 kilometers) toward the southeast and accounts for the higher altitude of the wine-growing region, which features elevations of 3,000–4,500 feet (914–1,371 meters). The area is bounded to the north by the southern slope of the Canadian River valley and on the east by the irregular and steeply cut Caprock Escarpment. The western boundary is the Mescalero Escarpment east of the Pecos River valley of New Mexico. The southern border is less defined as it gradually decreases in elevation to blend into the Edwards Plateau and the Johnson Creek branch of the Colorado River, east of Big Spring.

Caprock Escarpment. This "hard-pan" caliche layer underlying the Llano Estacado developed a few feet (less than a meter) below the ground as mineral-rich subsoil particles cemented together, forming a rock-like layer resistant to erosion. The Caprock Escarpment was formed by erosion between 1 and 2 million years ago and created a natural boundary line between the High Plains and the lower rolling plains of west Texas.

Edwards Plateau. The southernmost unit of the Great Plains that reaches into South Central Texas. The region includes a thin, erosional soil over hard Comanche limestone, which extends north to create the sub-layer of the High Plains. Its lack of deep soils has historically made the region less than ideal for farming crops, but well suited to grazing cattle, sheep, and goats. In recent decades, these poor, limestone-rich soils have been leveraged for successful vineyard plantings.

Llano Uplift. This unique geologic district lies along the northeastern edge of the Edwards Plateau. Despite its name, it is not an uplift at all, but the result of younger Comanche limestone and sedimentary sea bed eroding over time, unroofing older, pre-Cambrian rock. The inner portion is underlaid by granitic and metamorphosed rock, with Paleozoic sedimentary rock rimming these outcrops. Though structurally considered an uplift, the region is topographically a basin threaded by the Llano and Colorado rivers. When the Balcones Fault lifted during the Cenozoic era (about 66 million years ago), the Llano Uplift rose in elevation.

Enchanted Rock. One of the Llano Uplift's most recognized features is a large, pink dome known as Enchanted Rock. This granite formation rises 425 feet (129 meters) above the base elevation of the State Natural Park for which it is named. Its high point is 1,825 feet (556 meters) above sea level, and the entire dome covers 640 acres (259 hectares). Climbing the rock is like climbing the stairs of a 30- or 40-story building. Enchanted Rock is the second-largest batholith of its kind after Georgia's Stone Mountain, and its bald vastness can be seen from miles away. The indigenous Tonkawa tribe once believed the massive rock was the source of spiritual powers.

Balcones Escarpment. This geologic fault zone is several miles wide, consisting of several faults that form a curved line across Central Texas from Del Rio to the Red River. Around Del Rio, it is about 1,000 feet (304 meters) high, decreasing in elevation around San Antonio and Austin and dipping as low as about 300 feet (91 meters) high.

Texas Hill Country AVA

The Texas Hill Country AVA is geographically the third largest in the United States, and the largest overall in Texas. Spanning a total of 9 million acres (3.6 million hectares), the region is situated on the geological formation of the Edwards Plateau, a broad landscape consisting primarily of limestone bedrock. Its boundaries lie in central Texas, generally northwest of San Antonio and directly west of Austin. Its northern border is just north of San Saba, while its western edge is just west of Mason and Brady. Above the limestone bedrock, soils can be stony, gravelly, dark alkaline clay, and clay loam. The northern part of the Hill Country includes the Llano Uplift, a low geologic dome of igneous and metamorphic rock, primarily granite. Along the uplift are layers of sandstone, granite, schist, and iron-rich soil.

The region lies along the Balcones Fault, where it meets rolling prairie land to the east. The fault line produced a string of artesian springs, known as Comal Springs, that create the Comal River. Flowing for a mere 3 miles (5 kilometers) before entering the Guadalupe River, the Comal is considered the shortest river in the world.

The northwest part of the region is unique for the granite protrusions known as the Llano Uplift. This geologic wonder is not actually an uplift at all, but the result of younger sedimentary sea bed eroding over time, leaving older rock pre-dating the Precambrian era. When the Balcones Fault lifted during the Cenozoic period (more than 60 million years ago), the Llano Uplift rose in elevation.

The Hill Country is revered for its majestic rolling hills, steep canyons, and picturesque vistas. Elevations range from just 425 feet (130 meters) above sea level to more than 2,100 feet (640 meters). As such, many of the small towns throughout the region have become havens for destination travel. Drawing on the Hill Country's tourism success, many Texas producers have set up shop throughout the area. In fact, of the more than 400 permitted wineries in the state, nearly one-third have a tasting room within this AVA—a point of reference that will almost certainly be outdated by the time of publishing as new wineries continue to add to the ever-changing landscape of this region.

Though exact numbers vary, at recent count, there are an estimated 1,000 acres (404.6 hectares) of vineyards planted in the Hill Country. And there is growing interest from newer producers, such as Slate Mill Wine Collective, Lewis Wines, Southold Farm + Cellar, William Chris Vineyards, and Crowson Winery to plant significantly more. Historically, many of the vineyards planted in the Hill Country were both experimental and cosmetic. Many early producers wanted the look of lush vineyards to beautify their property and draw in more tasting room traffic. For many years, fruit for the wines produced by these wineries was often purchased in bulk from states such as California. The estate vineyards may have been used for some blending. But over the past decade, more producers have emphasized vineyard site selection and have taken greater care in researching soils, microclimate, aspects, and elevations. This is particularly the case in certain parts of the region that have uplifts and unique rock formations—such as around the Llano Uplift—as well as in areas with significant riparian influence from the Colorado River and its estuaries, where soils can vary from rocky and gravelly to silty and clay-based.

By and large, the growing season in the Hill Country begins in early March, though with the mild winters in central Texas, buds have been known to break as early as late February. The weather during fruit set generally remains dry with a fair amount of relative humidity and temperatures typically reach up to 98°F (36.6°C) during the day, falling to around 70°F (21°C) at night. However, it's not unusual to see temperatures reaching well over 100°F (37.7°C) for several consecutive days.

Texas High Plains AVA

Roughly 400 miles (644 kilometers) north of the Hill Country, the Texas High Plains AVA is only slightly smaller in size, covering about 8 million total acres (3.2 million hectares). In contrast to the picturesque, tourist-centric Hill Country, the High Plains is somewhat less visually stimulating, at least in terms of what the average wine traveler might expect. Located in the northwest part of the state known as "the Panhandle", the land is arid and desolate, with a persistent wafting of dust kicked up from the red alkali-rich caliche soils that cover a subsoil high in calcium a few feet beneath the surface. The landscape is flat, without a single visible rise, rock formation, or tree for miles. Historically, this vast expanse has been home to seas of cotton, peanut, and row-crop fields. But in recent years, those landscapes have been interrupted with the occasional flourish of vineyards. While many experts might take one look at this uninspiring, dust-covered region and conclude that it's not ideal for cultivating vines, you'll find more than three dozen grape growers who can prove those assumptions wrong.

What the High Plains may lack in relative beauty, it more than makes up for in substance. Indeed, while the Hill Country may have more tasting rooms and foot traffic, the High Plains has significantly more vineyard plantings. With more than an estimated 5,000 acres (2,023 hectares) under vine (more than two-thirds of the state's overall plantings), the region is the primary source of grapes for Texas wine producers.

With elevations between 3,000 and 5,000 feet (914 and 1,525 meters), the High Plains is a quadrangular, mesa-like area that rises perceptibly by sharp escarpments above the adjacent lowlands to the east and the west. The topography is characterized by thick deposits of wind-blown materials that blanket the region. The AVA is part of the Llano Estacado, a high mesa sloping at a rate of approximately ten feet per mile toward the southeast, which is one of the largest tablelands in North America. Its eastern border lines the 3,000-foot (914-meter) elevation contour along the Caprock Escarpment, a "hard-pan" layer that developed a few feet below the ground as highly mineral subsoil particles cemented themselves together to form a rock-like layer that resists erosion.

Reddy Vineyards, Texas High Plains AVA

The soil in the High Plains is deep, with almost universally dark-brown to reddish-brown sands, sandy loams, and clay loams, with some caliche outcroppings throughout. During the growing season, temperatures swing between above 100°F (37.7°C) in the day to below 60°F (15.5°C) at night, a broad enough diurnal shift to afford the grapes sufficient ripening.

Black Spanish

The Black Spanish grape has a long history in the Southern United States. The black-skinned grape variety has gone by many names, including Lenoir, Jacquez, Jack, Blue French, Ohio, and El Paso. While it has long been accepted as a native grape of Texas, the origins of Black Spanish remain murky. The National Grape Registry identifies its lineage as a crossing of *Vitis aestivalis*, *Vitis cinerea*, and *Vitis vinifera*. Its American wild grapevine parent is *Vitis berlandieri*, the same variety whose rootstock Thomas Munson sent to Europe to help solve the phylloxera dilemma in the late nineteenth century (see box, page 59).

There is evidence that the original parent of the variety made its way to the Madeira Islands sometime in the eighteenth century and was known as Jaqué or Jacquet. Noted South Carolina viticulturist Nicholas Herbemont recorded studies of the variety he named Lenoir after the man who gave him seeds he had cultivated near Stateburg, South Carolina. Black Spanish, Lenoir, and Jacquez were thought to be different varieties until the 1870s, when researchers discovered associations between the three that indicated they were identical. The only main differences were attributed to varying vintages and regional growing conditions.

Though also found in Alabama, California, Georgia, South Carolina, and Louisiana, Texas is the leading producer of Black Spanish, with more than 250 acres (101 hectares) in production in the eastern and southeastern parts of the state. The oldest documented planting dates back to the early 1880s when an Italian immigrant, Frank Qualia of Val Verde Winery, planted it in the southwestern town of Del Rio on the Mexican border.

Black Spanish is a moderate- to high-vigor grape variety known for its large, compact clusters with small berries. The inky red variety has been widely embraced in warmer, humid regions for its resistance to Pierce's disease, which is a common threat to *Vitis vinifera* in mild winter areas of the United States. It is commonly planted on its own roots and thrives in a wide range of soil pH, which is why it is planted throughout the state. Its ability to retain high titratable acidity in hot growing climates has made it a popular variety for blending.

Known in French as a *teinturier* variety, Black Spanish is one of the few red varieties whose juice is naturally pigmented, similar to Alicante Bouschet, Dunkelfelder, and Saperavi. The resulting wines are rich in color and concentration, with notes of jammy blackberry and spice. The grape's dark, juicy flesh yields wines that are sometimes musky on their own, but pleasurable when made into sweet, fortified wine with notes of cherry and chocolate.

> Aside from Val Verde Winery, which has championed the grape for more than a century, other notable Texas producers have helped to highlight the best aspects of Black Spanish. Raymond Haak of Haak Vineyards & Winery in Santa Fe, Texas produces "Port" as well as a Madeira-style wine called "Jacquez", using an in-house "estufa", or heated barrel-aging room. Messina Hof Winery in Bryan has been producing its flagship "Ports" from Black Spanish since the early 1980s. The "Paulo Porto" wines come in ruby, tawny, and signature Port-style offerings and are among the winery's best-sellers.

Other appellations

Located in far west Texas, the **Mesilla Valley AVA** is technically the oldest in the state. However, most of its 280,000 acres (113,312 hectares) are located in New Mexico, with only small parts extending into Texas. Named after an indigenous village christened Trenqel de la Mesilla by Spanish explorer Don Juan de Oñate, who came to the area in 1598, the region itself later became known as the Mesilla Valley.

Bell Mountain AVA has the distinction of becoming the first AVA located entirely within the state, earning the designation in 1986. Located in Gillespie County, west of Austin's Travis and Hays counties, Bell Mountain AVA is contained wholly within the larger Texas Hill Country AVA, covering about 3,200 acres (1,295 hectares) on the south and southwestern slopes of Bell Mountain. Elevations within this specific region range between 1,650 and 1,950 feet (503 and 594 meters). Several tributaries of the Colorado River, including the Llano and Pedernales rivers, cross the region west to east and join the Colorado as it cuts across the area to the southeast.

Located within the Texas Hill Country AVA, the **Fredericksburg in the Texas Hill Country AVA** specifically surrounds a large concentration of vineyards within the area of Fredericksburg, a small German-heritage town often thought of as the hub for Hill Country tourism.

The **Escondido Valley AVA** spans about 32,000 acres (12,949 hectares) in far west Texas near Fort Stockton.

Located in far west Texas, near the Mexican border, the **Texas Davis Mountains AVA** boasts the highest elevations in the state's wine-growing regions, ranging between 4,500 and 8,300 feet (1,371–2,530 meters).

Near the Texas–Oklahoma border, the **Texoma AVA** is notable for its essential role in the history of European viticulture. It was here in the late nineteenth century that viticulturist Thomas Volney Munson first grafted *Vitis vinifera* onto American rootstocks to help stop the spread of phylloxera, which had decimated the vineyards of Europe.

GRAPES

The early years of the Texas wine industry saw a flurry of plantings of Bordeaux varieties. (The exception was Doc McPherson's planting of Sangiovese. He believed the soils and climate of the High Plains were similar to those of the Mediterranean regions of Italy.) And while Cabernet Sauvignon remains one of the most widely planted varieties in the state, many producers have gravitated towards warm-climate varieties, looking to places such as Portugal, the Rhône Valley, and the warmer regions of Spain and Italy for direction.

In the late 1990s in particular, Texas began to see a departure from the classic varieties with the introduction of Tempranillo in the Hill Country by Jim and Karen Johnson of the former Alamosa Wine Cellars and in the High Plains with grape grower Neal Newsom under the encouragement of winemaker Dan Gatlin of Inwood Estates. Tempranillo not only has early ripening benefits but also displays hardiness against unpredictable weather. Around the same time, Viognier was beginning to take root with grape growers such as Bingham Family Vineyards making a serious investment in the variety by planting it in the High Plains in 2004. The white Rhône variety thrives particularly in the High Plains, ripening easily, and consistently demonstrating balanced acidity while producing desirable yields. In the vineyard, Viognier has proven to be both vigorous and resilient. Over the ensuing decade there was a significant increase in the success of these two varieties in particular.

Though Tempranillo and Viognier have enjoyed the spotlight over the past decade, there is no consensus among Texas producers that these two varieties are signature grapes for Texas. Indeed, while many of the state's top producers may include offerings in their portfolio, other varieties also have their champions, including Mourvèdre, Tannat, Touriga Nacional, Roussanne, Vermentino, Cinsault, Albariño and Picpoul Blanc.

In 2019, the United States Department of Agriculture conducted a survey of grape variety plantings throughout the state. The report found that 71 percent of grapes planted in Texas are red varieties and

29 percent are white. While Cabernet Sauvignon continues to be one of the most widely planted varieties, accounting for more than 700 acres (283 hectares), or about 12 percent of overall plantings, there was a significant uptick in plantings of Mourvèdre, which grew from 77 bearing acres (31 hectares) in 2015 to 250 (101 hectares) in 2019. Tempranillo is second to Cabernet Sauvignon with more than 450 total planted acres (182 hectares) with Merlot, Blanc Du Bois, Mourvèdre, Black Spanish, Sangiovese, Cabernet Franc, Viognier, and Malbec rounding out the top ten most planted varieties.

Between 2015 and 2019, Texas experienced substantial growth in red variety plantings of Tannat, Alicante Bouschet, Touriga Nacional, Petite Sirah, and Cinsault. In the white variety category, Blanc Du Bois slipped into the top spot for overall plantings, with a total of nearly 300 acres (121 hectares). Albariño made its first appearance in the study with a total of 60 acres (24 hectares) planted, while varieties such as Roussanne, Marsanne, Vermentino, and Trebbiano each saw an increase in overall plantings. It's important to note that this study is only an estimated snapshot of the overall plantings in Texas as there was less than 100 percent participation in the inquiry. To date, grape growers are not required to certify their vineyard or crop. There is still much experimentation taking place throughout the state, with different varieties being planted each year. Even so, the prevalence of warm-climate varieties in just the past few years reveals a clear pattern among growers to hone in on those cultivars that will work best in the state.

CHALLENGES

Considering the full range of growing conditions, the top growing challenges in Texas include hail, late spring frost, and late summer rains. But winegrowers have quickly adapted to these conditions, taking notes on the lessons learned in other growing regions across the country, particularly California and Washington.

Striking a balance with acidity, pH, and ripeness is a challenge because of the heat. Still, those winemakers who have become proficient in their processes are walking the tightrope more nimbly and would argue that this is simply one component of Texas terroir. Much of it means throwing the conventional wisdom of what may work in California out of the window in exchange for focusing more on what vines experience in Texas soils.

Blanc Du Bois

Though its name may lead many to assume this variety hails from France, Blanc Du Bois is actually a white hybrid grape that has primarily been planted throughout the southeastern United States. It was developed in 1968 by Texan John Mortensen at the University of Florida as a crossing between Florida D 6-148, and Cardinal, a California table grape. The name Blanc Du Bois—with both capital "D" and "B"—was given in honor of Emile DuBois, a French viticulturist and winemaker who migrated to Tallahassee, Florida, in the early 1880s.

In warmer, more humid regions, Blanc Du Bois grows well, showing resistance to Pierce's disease and downy mildew. And though its origins may be in Florida, its success for more than three decades in the Lone Star State has led Texas to adopt it as one of its "native" grapes. According to a 2019 report by the United States Department of Agriculture, Blanc Du Bois has overtaken other white varieties such as Muscat and Viognier as the most planted white grape in Texas, accounting for 15 percent of total white grape plantings. More than 40 percent of it is planted in the more humid parts of the state to the east and southwest, closer to the Gulf of Mexico.

Blanc Du Bois is a moderate to highly vigorous grape variety known to be an early ripener. Bud break typically happens in mid-March with harvest sometimes beginning in early July in the areas closer to the Gulf Coast. Offering distinctive floral aromas with notes of tropical fruit, citrus, honey, and stone fruit, Blanc Du Bois can produce refreshing, crisp dry and sweet styles of wine. Producers such as Haak Vineyards & Winery have been early champions of the grape, growing it in their estate vineyards near the steamy Gulf Coast in Santa Fe, Texas. Haak is the largest producer of Blanc Du Bois in Texas, offering a range of balanced dry, semi-sweet, and sweet table wines. Haak has also garnered a reputation for Madeira-style dessert wine made from Blanc Du Bois as well as Jacquez (also known as Lenoir or Black Spanish, see box, page 76).

The viticultural challenges within the state are many, particularly in the Hill Country and in the more humid regions, where mildew, fungus, nematodes, cotton root rot, and Pierce's disease are constant threats. The High Plains manages to avoid most of these maladies due to its higher elevation, dry climate, and ample wind activity, keeping vineyards dry and generally disease-free. In recent years, the vineyards have seen a threat from pesticide applications to nearby cotton and row-crop plantings. But with each passing vintage, growers are looking to see how the vines adapt.

While hail and wind can be detrimental to vineyards, late spring frosts are perhaps the most significant problem in Texas, particularly in the High Plains, where frosts have occurred as late as April and early May.

One crucial challenge to note is the somewhat unusual geography of winery production facilities in relation to the majority of the state's vineyards. Though wineries are scattered throughout the state, very few of them are actually located in the High Plains, where most of the state's fruit is grown (more than 85 percent). Instead, the majority of the harvest is trucked more than 400 miles (644 kilometers) from the High Plains south to wineries in the Hill Country, where tourism is much higher, or elsewhere in the state. This scenario creates a particularly distinctive relationship between grape growers and winemakers, many of whom spend little time seeing their grapes during the growing season. Instead, they rely on trust and communication with grape growers to farm their grapes. While it does seem unusual, it's worth noting that Washington state is perhaps one of the most successful American models of this. Most grapes are grown in the far eastern part of the state, while most wineries and tasting rooms are in the western part of the state, closer to Seattle. It's also the same model in Colorado, Arizona, and New Mexico, each having a region where grapes grow well that is far removed from the tourist areas, where most wine sales take place.

Still, some would argue that the state's most significant barrier to success is the lack of an overall industry commitment to making wines with Texas-grown fruit. Much of the Texas wine sold includes a percentage of wine made from out-of-state grapes. It's a common practice among many states with smaller wine industries and falls within the federal labeling guidelines. But those who want to see Texas press forward feel the labeling laws need to be much more stringent and have advocated for stricter guidelines to be enforced by the state government.

7

TEXAS PRODUCERS

TEXAS HILL COUNTRY AVA

Ab Astris

Stonewall
www.abastriswinery.com

One of the newer family operations to join the Texas Hill Country scene, Ab Astris was founded in 2018 by Tony and Erin Smith. Their daughters, Kathryn and Kristen, along with their husbands, Mitchell Sharrock and Michael Nelson, also moved to the region to join the family endeavor. Nelson pivoted from his career as an attorney to adopt winemaking as his new profession.

Located just along the Pedernales River, the 106-acre (42.9-hectare) property lies in the river basin with about 12–24 inches (30–60 centimeters) of sandy loam layered on top of limestone bedrock. The family manages 12 acres (4.8 hectares) of estate vineyard, which is planted to Tannat, Souzão, Clairette Blanche, Petite Sirah, and Montepulciano. In addition, Ab Astris sources fruit from a separate Hill Country-based Hoover Valley Vineyard further north as well as from the High Plains from Newsom Vineyards, Reddy Vineyards, and Narra Vineyards. With a motto of "all Texas, all the time", Ab Astris has shown a commitment to featuring the grape varieties that grow well in the regional soils of both the Hill Country and the High Plains. While it does produce some blends, its strength is in showcasing the hard work done in the vineyards from year to year with vineyard-designated releases.

Inspired by a quote from famed Italian astronomer Galileo, "wine is sunlight held together by water", the winery name is Latin for "of

82   THE WINES OF SOUTHWEST U.S.A.

the stars", with each label bearing a celestial design. Of particular note are the beautifully structured and elegant Tannat, the lithe and crisp Picpoul Blanc, and the tart red fruit and earthy undertones of the Montepulciano.

Becker Vineyards
Fredericksburg
www.beckervineyards.com

Richard and Bunny Becker are considered foundational visionaries in the Texas wine industry, with their property being one of the early Texas Hill Country wineries to establish a presence in the region. Dr. Becker, an endocrinologist practicing in San Antonio, and his wife Bunny, a speech therapist, had traveled extensively to wine regions around the world and found many similarities in these far-off growing regions to the rolling hills of Texas. In 1992 they decided to plant wine grapes on a 46-acre (18.6-hectare) plot of land in Fredericksburg. They released their first vintage in 1995 before opening a tasting room in 1996. Over time, the property has evolved to include an event space, a larger tasting room, and 56 acres (22.6 hectares) of vineyard.

The Beckers have long bought grapes from Hill Country grower Drew Tallent, who supplies more than 80 percent of his harvest to them from his 65-acre (26.3-hectare) vineyard in Mason County. They also purchase grapes from a few grape growers in the High Plains, including Farmhouse Vineyards, Reddy Vineyards, Canada Family Vineyards, and Wilmeth Vineyards. Winemaker Jonathan Leahy executes a wine program geared towards bigger, riper styles of wine, allowing growers to let the fruit hang longer during harvest season to push for a richer flavor. The Beckers employ methods such as *délestage* (or rack and return) in the cellar and use new French oak for fermentation and aging white wines, with new American oak used for red wines, all of which help to define the overall Becker style of wine. Bordeaux and Rhône blends lead the Becker Vineyards portfolio of nearly two dozen wines. Of note are rich and luscious Cabernet Franc Reserve, the Prairie Rotie, a Rhône-inspired blend of Mourvèdre and Carignan with a touch of Merlot and Barbera, and the soft, broad palate of the Viognier Reserve.

Becker wines have been served to three presidents, at three governor's dinners at the White House, and at ten James Beard House Dinners in New York, not to mention the many food and wine festivals throughout Texas.

TEXAS PRODUCERS 83

Bending Branch Winery
Comfort
bendingbranchwinery.com

While it is true that Bending Branch Winery has made a name for itself by producing quality wines for the past decade, it's equally true that it has leveraged technology and modern winemaking techniques to push forth greater concentration, flavor, and structure in the wine program.

Founded in 2009 by Dr. Robert W. and Brenda Young, Bending Branch Winery is located just outside of the small Hill Country town of Comfort (about an hour northwest of San Antonio). It might be easy to suggest that Young, like many doctors and lawyers before him, casually waded into the wine industry for a more relaxed, romantic second career. But the notion could not be further from the truth. To be sure, winemaking is the former physician's second career, but there's nothing casual about the way he embarked upon the venture. Ever the academic, Young completed his enology studies at the University of California, Davis and immersed himself in researching the right grapes to consider for his Hill Country wine program.

He narrowed down varieties that would thrive best in hot climates and poor soils, and that had late-spring bud break and a later harvest date to help mitigate the threat of late spring frosts. He wanted to work with varieties that would maintain natural acidity and be relatively resistant to drought. Finally, Young sought grapes that were high in procyanidins, the most active of the polyphenols—compounds proven to lower the risk of both heart disease and cancer. After completing his research, Young selected Tannat, Souzão, Picpoul Blanc, Charbono, and Petite Sirah as a few of his top picks. (Young was one of the first in Texas to begin working with Picpoul Blanc and Tannat, an endeavor that has met with great success.) He developed relationships with growers in the High Plains and Hill Country to grow these varieties. He also planted a small estate vineyard with Crimson Cabernet, a genetic cross between Cabernet Sauvignon and Norton that is resistant to Pierce's disease.

In the cellar, the Bending Branch wine program focuses primarily on big and bold reds using small-batch, hand-crafted open-top fermentation and innovative fermentation processes, including cryo-maceration and thermovinification, or *flash détente*. To date, Bending Branch is the only producer in Texas to use the *flash détente* to extract more polyphenols and proanthocyanidins in red wines. Ridding the wines of unwanted tannins, but promoting the desired ones, Young has made a name for himself in crafting wines that have become recognized for structure and quality.

To capitalize on the growing amount of foot traffic on Comfort's High Street, the town's main thoroughfare, the winery opened a second tasting room, Branch On High. Bending Branch's flagship wine is Tannat, which has become a consistent show-stopper from vintage to vintage. Also noteworthy is Cabernet Sauvignon and the rich-yet-summery Tannat rosé.

Calais Winery

Hye
www.calaiswinery.com

You know there's potential for Texas wine when a Frenchman ditches his career in engineering to open up his own winery using Texas-grown grapes. Such is the story of Benjamin Calais, a native of Calais, France who found himself working a corporate gig in Dallas more than a decade ago. A hobbyist home winemaker, Calais' passion for the pastime led him to launch his own label in 2008 in the hip, up-and-coming neighborhood of Deep Ellum. It would take a few years before he decided to leave Dallas for the Texas Hill Country, where he has expanded into a better winemaking facility.

Focused solely on Bordeaux-style blends at first, Calais forged relationships with grape growers in the High Plains and settled on his winemaking regimen, which includes letting the wine have as much time as needed to show at its very best. Calais works with Cabernet Sauvignon, Merlot, Cabernet Franc, Petit Verdot, Sauvignon Blanc, and Semillon. He uses 100 percent Texas grapes grown from high-elevation sites (between 3,200 and 5,400 feet; 975–1,646 meters) with enough time on the vine to reach full phenolic ripeness. In production, red and white wines are fermented using both wild and commercial yeast depending on the specific wine. Wines spend time in 225-liter French oak barriques. With a strong emphasis on red varietals, Calais Winery is known for its complex wines displaying concentration and body. If available, try the Cuvée L'Exposition Cabernet Sauvignon from Newsom Vineyards with notes of summer cherries, savory tomato, and cedar, or the Cuvée du Rocher, a true Bordeaux-style Sauvignon Blanc balancing vibrance and broad structural appeal.

Crowson Winery & Tasting Room

Johnson City
crowsonwines.com

Henry Crowson worked as a cellar rat and assistant winemaker at William Chris Vineyards before becoming a notable player in the Texas wine story. Inspired by the natural winemaking methods of California's Tony Coturri and Lewis Dickson of La Cruz de Comal, Crowson dug deep into understanding the best way to bring a voice to the fruit from Texas vineyards in the most authentic way possible, before launching his own brand in 2018. The winery is a true boot-strap operation that began as a shared experience with his one employee, his father, Kenny Crowson. The younger Crowson initially leveraged his relationships with growers in the High Plains to source his fruit and today works with Narra Vineyards and HossFly Vineyards, as well as a few others when fruit is available, to fulfill his small production of about 2,000 cases.

Crowson favors single variety releases over blends to allow for a unique expression of regional terroir. As is expected with most natural wines, Crowson wines are fermented with native yeasts and are unsulfured, unfined, and unfiltered. Try the vibrant and aromatic Malvasia Bianca, the alluring Montepulciano, or the fruity and chewy Sangiovese rosé.

Duchman Family Winery

Driftwood
duchmanwinery.com

Easily one of the state's most consistent producers, Duchman Family Winery entered the Texas wine scene in 2006, championing Italian grape varieties long before anyone had given them a serious look. Owned by Dr. Stan and Lisa Duchman (pronounced *duke-man*) from Houston, the winery has made its home in Driftwood, a few miles west of Austin in the Hill Country. Winemaker Dave Reilly came into his role as head winemaker under the tutelage of his friend and mentor Mark Penna, who had previously served as the winemaker for Llano Estacado, Pheasant Ridge, CapRock Winery and Ste. Genevieve Winery. Penna taught Reilly everything he knew and handed over the reins in 2008. Since then, Reilly has made his mark with varieties such as Montepulciano, Vermentino, Trebbiano, and Aglianico.

Reilly sources the majority of his fruit from the High Plains, from growers including Bingham Family Vineyards, Oswald Vineyards, and

86 THE WINES OF SOUTHWEST U.S.A.

Reddy Vineyards. In the Hill Country, he sources from the Salt Lick Vineyards and Hog Heaven Vineyards. Duchman wines are consistently true to the character of their variety and are at home on any dinner table. In the cellar, Reilly produces elegant red wines aged mostly in neutral oak barrels. White wines are exclusively produced in stainless steel tanks. Notable wines include the Vermentino, a flagship white grape for Duchman that sings when enjoyed with the Texas heat, and the bold and rustic Aglianico, a variety truly at home in the baked Texas soils.

Fall Creek Vineyards
Tow
www.fcv.com

For more than 40 years, Ed Auler has been a firm believer in the potential for Texas wines. He and his wife, Susan, were just behind Bob Reed and Doc McPherson in pioneering the Texas wine industry, with the creation of Fall Creek Vineyards near Lake Buchanon in the late 1970s. The Aulers have since been among the most influential voices in the Texas wine story.

It all began with a simple trip to France in 1973. As part of a five-generation cattle ranching family, Auler had planned a trip to learn more about the French cattle breeds, and decided to visit the wine regions of Champagne and Bordeaux along the way. Inspired by a copy of *A Wine Tour of France*, by Frederick S. Wildman, the trip took a turn from exploring more about cattle to learning more about wine. As the Aulers recall, they spent three days learning about cattle and three weeks learning about wine. Ed Auler was struck by the similarities between the French Rhône countryside they were driving through and his family's ranch in the Texas Hill Country.

As misfortune would have it, the bottom dropped out of the cattle market in 1974, and the Aulers quickly turned to the idea of using their land to create something else, and growing grapes. In 1975 they planted a test plot with nine French/American hybrids, including Emerald Riesling and Ruby Cabernet and two *Vitis vinifera*, Chenin Blanc and Cabernet Sauvignon. The results within the first few years were promising. At the time, wine growing in Texas was for adventurers. Nobody knew what to do or what to plant. Slowly, they began to add other varieties, including Merlot and Chardonnay.

With the help of consultants from the University of Texas, Texas A&M University, Texas Tech, and Dr. Vincent Petrucci of Fresno State

University—an early believer in Texas wine grapes—the Aulers steadily began to get their bearings with viticulture. By 1983, construction of their winery was complete, and Fall Creek Vineyards had a few vintages under its belt. That same year, Fall Creek's first Chenin Blanc topped the *Wines and Spirits* magazine annual Buying Guide as the best Chenin Blanc in all of America for that vintage.

The Aulers also sought and received help from famed Russian winemaker, André Tchelistcheff, who is known for putting California wines on the map, and consulting with the likes of Robert Mondavi and Louis Martini. Upon tasting the 1985 vintage, which was the most substantial effort the winery had yet put forth, Tchelistcheff insisted the Aulers plant more grapes.

In many ways, the Aulers have dedicated as much time to building the Texas industry as they have their own brand. Drawing on his legal background, Ed Auler spearheaded the effort to create the Texas Hill Country American Viticulture Area (AVA) with the Bureau of Alcohol, Tobacco, and Firearms (BATF), receiving the designation in 1990. In 1985, Susan leveraged her contacts within the restaurant industry around the state to launch the Texas Hill Country Wine and Food Festival. This event would run for more than two decades before being acquired by the Austin Food & Wine Festival in 2012.

In the early years, the Aulers started with just a few hundred cases of wine, which they would distribute themselves from the trunk of their car. But over four decades, the Fall Creek brand has grown to produce more than 50,000 cases annually. For the majority of that time, Ed was the primary winemaker. But in celebration of their thirtieth year, the Aulers sought to affect a significant change for their life's work and brought on a new winemaker, Sergio Cuadra. Hailing from Chile, Cuadra was the personal recommendation of noted Sonoma producer Paul Hobbs. He had spent 19 years in the Chilean wine industry with notable Bordeaux wine expert Jacques Lurton at Viña San Pedro, Master of Wine Kim Milne at Caliterra, and most recently at Anakena, where Hobbs worked as a consultant. Before that, Cuadra spent a decade at Concha y Toro as one of the principal winemakers.

Excited at the opportunity to be a part of an emerging wine region, Cuadra moved his wife and five children to the Hill Country to start a new life as a Texas winemaker for the 2013 vintage. The addition of Cuadra was well worth it. His experience with some of Chile's warmer microclimates allowed him to approach the Texas vineyards with fresh eyes. While Auler learned a whole career's worth of new methods in the winery, the experience entailed a learning curve for Cuadra as well. In addition to the estate vineyards at Fall Creek's winery in Tow, the winery sources grapes from a few places throughout Texas, including Voca in the western part of the Hill Country, Mesa in west Texas, and Driftwood, near Austin. Cuadra was surprised at how adaptable the vines can be to different conditions and still produce quality grapes. To him, it made no sense for Sauvignon Blanc, typically a cool-climate variety, to do well in Texas. But he soon learned that temperature was not the only factor for a variety such as this. Soils and growing conditions were also crucial for the plant to adapt. For Cuadra, coming to Texas has allowed him to see broader possibilities for cultivating different varieties.

While Fall Creek started with Bordeaux varieties, along with Chardonnay and Chenin Blanc, its evolution over the years has broadened its scope to include more warm-climate varieties such as Tempranillo, Grenache, Syrah, and Mourvèdre. If available, try the latest release of Meritus, a blend of red Bordeaux varieties, or the pricier, yet stunning single variety offerings under their ExTerra label (the Syrah and the Tempranillo are particularly exceptional).

French Connection Wines

Hye

www.frenchconnectionhye.com

The second brand from French winemaker Benjamin Calais, the French Connection is a collaborative project including business partners Sheri and Pat Pattillo, and Victoria Gimino. While Calais had already established a Bordeaux program with Calais Winery, he was eager to launch a separate Rhône program independent of his first venture. In 2019, the right opportunity came along, and he formed a partnership with the Pattillos and Gimino.

French Connection Wines purchases fruit from vineyards in the High Plains and specializes in making wines specifically from Rhône varieties, including Syrah, Mourvèdre, Cinsault, Grenache, Counoise, Carignan, Roussanne, Marsanne, Viognier, and Picpoul Blanc. In the cellar, all red wines spend time in mostly French oak puncheons, while white and rosé wines see French oak barriques or stainless steel. Be sure to try the rich yet refreshing La Connection Rosé (100 percent Mourvèdre) or the peppery and robust La Connection Red, a fifty-fifty blend of Mourvèdre and Syrah.

Hye Meadow Winery

Hye

hyemeadow.com

Having fostered a love for Italian wines from an early age, Mike Batek had a feeling there would be a time in his life when wine might take a central role. When he and his wife, Denise, hit a crossroads in their careers in Austin in 2012, that early intuition became a reality. Settling on an idyllic piece of land in Hye, about midway between Johnson City and Fredericksburg, the Bateks launched Hye Meadow Winery, planting 7 acres (2.8 hectares) of estate vines to Aglianico, Montepulciano, Negroamaro, Refosco, and Tempranillo. The estate vineyards support about 25 percent of their wine production, with the Bateks sourcing the remaining 75 percent from vineyards in the High Plains. Hye Meadow red wines are typically destemmed into small bins and punched down several times a day, followed by barrel fermentation and aging. White wines are generally fermented in stainless steel tanks and rest on the lees for several months. Try the sunny and vibrant Viognier with a soft palate and notes of apricot and summer flowers, or the Full Monte, a Montepulciano with notes of jammy black plum and an earthy finish.

The Infinite Monkey Theorem

Austin

theinfinitemonkeytheorem.com

The one winery that is considered both a Texas and Colorado winery, the Infinite Monkey Theorem (IMT) began in Denver in 2008 as the brainchild of husband-and-wife team, Aaron and Meredith Berman. The Bermans had just completed MBA degrees at the University of Colorado and were looking for a way to marry their creative spirit with a proven business model. But instead of building something traditional—say, an idyllic vineyard in rolling countryside—they opted for something less conventional: an outfit right in the heart of Denver. They began by sourcing grapes from western Colorado—where there's a bounty of farmland—trucking them over the Great Divide of the Rocky Mountains and crushing them in Denver.

The name is a reference to the theory that suggests that given an infinite period of time, a hypothetical monkey, tapping keys at random on a typewriter, is likely to type out a full text, perhaps even the complete works of Shakespeare. All of the wine bottles have silk-screened labels—a first in the wine industry—and the bold design is of a chimpanzee glaring out in front of a spray of graffiti art in the background. The team was the second wine producer in the United States to package their wines in cans—long before canned wine became popular.

In 2015 they turned their attention to Texas, where they duplicated the urban winery concept in Austin. They first reached out to celebrated Texas winemaker Kim McPherson, of McPherson Cellars in Lubbock, to help get their first Texas wine produced while they built their own production facility. Today IMT's winemaking team churns out a nationally distributed line of canned American wine along with a selection of Texas wines made from grapes grown in the High Plains. Try the leathery, concentrated Tempranillo, the golden, peachy Viognier, and if you can find it, the 2016 Merlot, an exceptional offering of this variety, which has rarely been a star performer for Texas.

Kuhlman Cellars
Stonewall
kuhlmancellars.com

One of the newer producers in the Hill Country, Kuhlman Cellars seamlessly launched into the world of Texas wine in 2014, offering pristine wines and generous hospitality. For owners Jennifer and Chris Cobb, along with Chris' parents Diane and Reed Cobb, the effort was undoubtedly helped by their choice of winemaker, Bénédicte Rhyne. Originally from France, Rhyne brought with her an established reputation in the Texas wine community, having made wines under the broad umbrella of Ste. Genevieve for many years. She began her career in Sonoma with Ravenswood Winery, after completing her Diplôme National d'Œnologue (Masters in Enology) at the University of Burgundy in Dijon, France. In search of a more boutique winemaking experience in Texas, Rhyne joined the team in 2012 as a winemaker and partner.

Kuhlman Cellars took its name from Kuhlman Creek, which originates on the Cobb family farm and ultimately empties into the Pedernales River. The Cobb family owns and manages three vineyards within the Hill Country, but also sources grapes from both the High Plains and Escondido Valley. From day one, Rhyne's wines have revealed she has a deft hand at letting the regional influences of the High Plains and the Hill Country shine through. Notable selections include the Alluvé red blend of Mourvèdre, Tempranillo, Carignan, Grenache, Malbec, and Petite Sirah as well as the delicate Cinsault Rosé and the Sauvignon Blanc, which manages to shine in the sandy loam soils of west Texas.

La Cruz de Comal

New Braunfels
lacruzdecomalwines.com

Perhaps one of the state's most eclectic and craft-committed producers, La Cruz de Comal is the culmination of what owner and winemaker Lewis Dickson believes authentic Texas wine should be. Having spent the majority of his career as one of Houston's most well-regarded trial lawyers, Dickson spent a few years living in Provence in the late 1990s. While there, he was struck by the similarities the terrain bore to the limestone soils and natural landscape of his Hill Country ranch near Canyon Lake (between Austin and San Antonio). In 2000, Dickson decided to plant an estate vineyard on his property, selecting 11 different varieties, including some *Vitis vinifera* as well as Black Spanish and Blanc Du Bois. After seven years of trial and error in the vineyard, Dickson had determined a few things; most significantly he had become committed to farming organically. With this in mind, he narrowed his varieties down to Blanc Du Bois and Black Spanish, determining that they were the best equipped to thrive in the microclimate of the vineyard and therefore the most suited to organic farming. Today, his vineyard is focused equally on these two varieties.

When it came to making wine, Dickson already had an idea of what he wanted to achieve. Since 1981, he had been visiting Sonoma's Coturri Winery and was drawn to the particular style of wines made by the proprietor, Tony Coturri. But at the time, he wasn't quite sure what made them so different. All he knew was they were unlike any others he'd ever tasted.

In 2001, Dickson prevailed upon Coturri to come to Texas and teach him his winemaking methods using grapes sourced from Robert Clay Vineyards in Mason County. This was when he discovered that Coturri had long employed the philosophy of low intervention winemaking methods in the cellar, allowing his wines to ferment using wild yeasts, with no sulfur, sugar, acid, or coloring agent additions, and no fining or filtering. Although today these winemaking practices fall under the fad category of "natural wines" Dickson, and Coturri for that matter, was drawn to this style of wine long before it was trendy.

For Dickson, what matters most is that the wines taste true to what the fruit naturally expresses, rather than subscribing to a specific method of winemaking simply to fit into a particular category. Today, Dickson remains faithful to what Coturri taught him in the cellar. While he

92 THE WINES OF SOUTHWEST U.S.A.

relies greatly on his own estate vineyard, he also purchases fruit from a few grape growers, specifically in the Hill Country, emphasizing a commitment to the regionality of his wines.

To say Dickson's wines are artisanal is an understatement. Everything is done by hand, from harvest to bottle. But he wouldn't change a thing. Dickson strives for a more Old-World style with acid-driven wines that reveal the vibrant fruit and floral characters of Blanc Du Bois, Black Spanish, and the handful of other varieties with which he works.

La Cruz de Comal is named after the Mexican cross that stands in the estate vineyard, an artful relic made from the pieces of *comals,* or *comalitas,* the iron pan used for cooking throughout Mexico. In Spanish, the cross made of comals translates to "La Cruz de Comal". Each vintage, Dickson's wines are anything but boring, echoing a clear voice of what the Texas soils and climate have to offer. Notable selections include the dry white Pétard Blanc (made from Blanc Du Bois), a Black Spanish red table wine called Troubadour, and the Quinta La Cruz, a fortified, red dessert wine made from 100 percent estate-grown Black Spanish. Also worth seeking out is the Après fortified dessert wine from Blanc Du Bois, the vermouth-style Shango Buen Comienzo, the funky-yet-fruity Frimouse Naturelle (a *pétillant naturel*), and the Madeira-style Falstaff's Sack.

Lewis Wines
Johnson City
www.lewiswines.com

Lewis Wines is particularly interesting in that it is one of the first wineries to see two young upstarts forgo a post-college career in something else before turning to wine. Both Doug Lewis, for whom the winery is named, and business partner Duncan McNabb were in their mid-20s when they launched Lewis Wines. Lewis owes his early passion for Texas wine to a fruitful internship at Pedernales Cellars under the direction of winemaker David Kuhlken. He and McNabb, a chemistry major who had played with Lewis in a college soccer league, opted to go all-in with the Texas industry in 2010. They began by making wines on the side while working at Pedernales Cellars, using grapes they had sourced primarily from the Hill Country. By 2011, they were ready to set out on their own. Their bootstrap operation started small and was carried through by a whole lot of hope. But the result has certainly been fruitful.

Lewis and McNabb manage a 10-acre (4-hectare) estate vineyard, named Five Lions Vineyard, planted with Touriga Nacional, Tinta

Cão, Alicante Bouschet, Tannat, and Arinto. They also help manage a 6-acre (2.4 hectare) leased vineyard, Round Mountain Vineyard, which supplies about a third of the winery's production. The remaining fruit comes from Hill Country vineyards, including Parr Vineyards, Robert Clay Vineyards, Klein Vineyards, Tio Pancho/Bear Vineyards, Enchanted Rock Vineyards, and Tallent Vineyards. They also work with High Plains growers, including Phillips Vineyard, Lost Draw Vineyard, Newsom Vineyards, and La Pradera Vineyards. The balance of wine produced from each region is about fifty-fifty, which speaks to an overall goal of producing regionally expressive wines. Rather than making a sweeping statement about star varieties to work with, Lewis sees each new year as a learning opportunity to figure out why they may prefer a particular variety in one given site, but not necessarily in another. Their hope with each growing season is to learn more about the terroir of a given AVA and vineyard site, and develop their farming practices, in order to understand how it all relates to the wines. As they continue to experiment with varieties, a few that have held their interest include Touriga Nacional (they are currently the largest producer of the Portuguese cultivar in the state), Mourvèdre, and Tempranillo.

While the first releases were primarily red wines of notable quality, including a Round Mountain Estate red blend of Touriga Nacional, Tinta Cão, and Tempranillo, Lewis Wines is also excelling at white and rosé offerings. The weighty yet crisp Estate Rosé is a consistent pleaser, as is the steely Chenin Blanc.

Lost Draw Cellars
Fredericksburg
lostdrawcellars.com

Lost Draw Cellars is the result of a best-case scenario. The winery is named after Lost Draw Vineyards, both a vineyard and grape consulting operation owned by Andy Timmons, one of the foremost grape growers in the High Plains. A multi-generational row cropper, Timmons transitioned to vines in 2005 in search of a more sustainable crop than peanuts and cotton. Employing an aggressive approach to learning about the best viticultural practices in the state, Timmons has sought the guidance and advice of both grape growers and winemakers in Texas, Washington state, and beyond. He was one of the first to bring in wind machines to help mitigate late spring frosts. From his extensive travels to other wine regions, he has learned best practices for canopy management, hail

94 THE WINES OF SOUTHWEST U.S.A.

protection, and pruning. Today Timmons is one of the largest farmers of grapes in Texas, managing more than 500 acres (202 hectares) and more than 30 varieties, including the 60-acre (24.3-hectare) Lost Draw Vineyards, the 100-acre (40.4-hectare) La Pradera Vineyards, and the 20-acre (8.1-hectare) Timmons Estate Vineyard. Perhaps key to Timmons' success has been his willingness to mentor and advise other growers in the High Plains.

Timmons, like many fellow growers, expanded his farming career into the realm of winemaking. He partnered with his nephew, Andrew Sides, an engineering graduate from Texas Tech University who had worked at the family farm throughout high school and college, and Sides' father-in-law, Troy Ottmers. The Ottmers family has a long history in the Fredericksburg area of the Hill Country. The Lost Draw winery and tasting room are located on a property near downtown Fredericksburg that has been in the Ottmers family since 1936. In 2012, Lost Draw Cellars produced its first vintage, opening the tasting room in 2014.

Today, winemaker Brad Buckelew works alongside Andrew Sides to produce a portfolio of wines that seems to increase in quality and intrigue with each passing vintage. More than 80 percent of Lost Draw production comes from the family vineyards in the High Plains (as well as a couple of others), with Hill Country fruit from a few smaller growers making up the difference. Buckelew and Sides use a low-impact winemaking approach to allow the wines to represent their terroir best.

Try the Reserve Sangiovese, a beautifully complex wine showing red fruit, earthy character, and elegant structure, or the Counoise Rosé, a delightful Provence-style rosé with notes of summer berries, and light florality.

Pedernales Cellars
Stonewall
www.pedernalescellars.com

Pedernales Cellars is an example of what happens when a family comes together with a clear vision. Of course, that vision took a few years to come into focus. In 1995, Larry and Jeanine Kuhlken came to the Texas Hill Country to plant a vineyard. The two empty nesters, who first met more than 50 years ago when working together at NASA on the Apollo 11 mission, were ready to leave a life in the information technology industry for early retirement. Although their grown children had gone on to pursue their own endeavors, they rejoined their parents to help establish the new planting. Their daughter, Julie Kuhlken, finished her undergraduate and Master's studies in California at Stanford University before moving to London to complete her doctorate in philosophy at Middlesex University. David, their son, had finished his undergraduate studies at Rice University in Houston and had just completed his Master's in Business Administration at the University of Texas when the question of what to do next became a pressing subject.

As the siblings watched their parents roll their sleeves up to work their land and sell their grapes to other wineries, the idea to start their own winery began to take root. David was already eager to find a way to start his own company, and his sister was also keen on the idea. In 2005, the two formulated a business plan for a boutique winery focused on small-lot wines expressive of the Texas Hill Country. David went on to take enology classes through the University of California, Davis, and in 2006 the siblings produced their first vintage. In 2008, they opened their winery and tasting room.

Pedernales Cellars received its name from its nearby geography, which is shaped by the Pedernales River that runs through this part of the region. While the winery is located in Stonewall, a small town mid-way between Johnson City and Fredericksburg, the Kuhlken Vineyards are located about 11.5 miles (18.5 kilometers) north of Fredericksburg in the Bell Mountain AVA, the oldest AVA in Texas. (The vineyard was formed in 1986 by Bob Oberhelman of Bell Mountain Winery.) It is uniquely situated on a part of the Llano Uplift, a geological formation that brought old, infertile soil to the surface around 300 million years ago, including the degraded sandstone found in Kuhlken Vineyards.

The original planting in 1995, which is known as Block Zero, included Sauvignon Blanc, Chardonnay, Merlot, and Cabernet Sauvignon. In the beginning, these grapes were sold to other wineries, who were caught up in the demand for more well-known varieties. While the Merlot and Cabernet produced good quality grapes, the white varieties did not thrive and were eventually replaced with Mourvèdre. Block Zero now includes Cabernet Sauvignon, Merlot, Sangiovese, and Mourvèdre. Two new sections, Block One and Block Two, were planted in 2007. Block One includes more Mourvèdre and Tempranillo, while Block Two includes Touriga Nacional.

With a commitment to Texas fruit for its entire portfolio, Pedernales focuses on working with grapes that are well-suited to growing in the local climate. Not only does this philosophy make sense from a

96 THE WINES OF SOUTHWEST U.S.A.

viticultural perspective, but it also ensures that the techniques used in the cellar can remain minimal in an effort to honor the regional terroir of Texas.

Pedernales Cellars sources the majority of its fruit from the High Plains but also works with Hill Country and west Texas fruit. Wines such as Dolcetto, GSM Melange, Albariño, and Graciano are all noteworthy offerings; the winery has become renowned for its Tempranillo and Viognier, two varieties which have gained significant attention in Texas. While the Viognier is exclusively from the High Plains, Kuhlken offers a few different Tempranillo releases, including a High Plains and a Hill Country selection, as well as a blend of the two regions and a reserve version of this blend. Tasting these wines side by side offers a clearer picture of the regional nuances the grape takes on. In general, Pedernales wines are balanced, well-structured, and straddle a fence between rustic and sophisticated.

Perissos Vineyard and Winery
Burnet
perissosvineyards.com

Located in Burnet, in the northern part of the Hill Country, Perissos Vineyard is a lovely estate vineyard situated in a fertile spot in the Colorado River Basin. Owners Seth and Laura Martin first planted a section of their vineyards in 2005 and opened their tasting room in 2009. The couple live on their estate vineyard and farm with their five children. A former homebuilder, Seth designed and built the winery's elaborately trussed tasting room and event space. The Martins primarily work with Mediterranean cultivars and other varieties that thrive in warm climates such as Aglianico, Montepulciano, Mourvèdre, Roussanne, Tempranillo, and Viognier.

Roughly 70 percent of Perissos wines are made from grapes grown in their 16-acre (6.5-hectare) vineyard, with another 2 acres (0.8 hectares) slated to be planted in the coming year. The other 30 percent is sourced from the High Plains, from vineyards including Branded H Vineyards, Diamanté Doble Vineyard, Kubacak Vineyards, Lahey Vineyards, Nogalero Estate Vineyard, and One Way Vineyard.

Perissos wines lean towards a riper, more concentrated style. The Aglianico will appeal to those who love opulent fruit characteristics backed by ruddy tannin and broad structure. Try the luscious Racker's Blend, a powerful field blend of Aglianico, Tempranillo, Petite Sirah,

Malbec, and Syrah, or the rich and rustic Aglianico, a signature grape variety for this producer.

Ron Yates
Johnson City
ronyateswines.com

As a second concept from the owners of Spicewood Vineyards (see p. 100), Ron Yates is a departure from its sister winery in that it looks outward, sourcing fruit from key grape growers in the state, rather than focusing specifically on estate-grown wines. Co-owner Ron Yates, for whom the winery is named, and winemaker Todd Crowell work closely with growers in High Plains, Escondido Valley, and the Davis Mountains to manage fruit throughout the growing season that can best express the region in which it is grown. While they do source fruit from the Hill Country, the opportunity to diverge from the estate model of Spicewood Vineyards has allowed Yates and Crowell to explore more within the other Texas wine regions. In the cellar, the winemaking method is minimal as a general rule, but extra attention has gone to developing a well-researched barrel program that showcases oak from some of the world's finest coopers and forests, including those in France and the United States. Tempranillo offerings easily rival world-class Rioja, and in a good vintage, Cabernet Sauvignon is equally impressive. The Viognier is also worth a try. Rather than using stainless steel for fermentation as many Texas producers have done, this program employs both new and neutral oak fermentation and aging to yield Viognier with a broader palate that is still balanced and lively.

Slate Mill Wine Collective
Fredericksburg
slatemillwinecollective.com

The first full-service winery incubator in the state of Texas, Slate Mill Wine Collective is already making a significant impact in the industry since its opening in early 2020. The vision of a west Texas oil family by the name of Jones, Slate Mill first began with the purchase of an existing winery and vineyard, 1851 Vineyards. In 2019, the Joneses took over the Fredericksburg property and started a significant expansion of the existing winery and estate vineyard planting. While the 1851 brand still lives on under the new venture, Slate Mill's core mission is to function as a custom crush facility to further the Texas wine industry by helping new, quality-minded producers get off the ground.

The wine program is led by veteran Texas winemaker Josh Fritsche, who oversees wine production of 1851 Vineyards but also works as a consultant for custom crush clients who have use of the facility. When working with Slate Mill Wine Collective, clients have the option to either handle all wine production or use the Collective's winemaking team for a full-service approach that extends to assistance with vineyard planting and management, as well as winemaking and sales, including wholesale and distribution, wine club shipping and fulfillment.

Slate Mill's typical clients have already made great strides in developing their brands, but have often been dependent on the generosity of other wineries during harvest season to allow them space to make their wines. But with this new collective, the anxiety over whether or not space will be available has disappeared. Would-be clients interested in producing under the Slate Mill umbrella have to apply and meet a set of standards put forth by the winery. Among the primary requirements for joining the collective are that wines must be made using 100 percent Texas fruit.

Current clients producing at Slate Mill make up a distinguished list of highly regarded upstarts including Randy Hester, winemaker for the C. L. Butaud label, who gained his winemaking chops in Napa Valley; Rae Wilson, winemaker for Dandy Rosé who spent much of her career as a sommelier and winemaking assistant; Farmhouse Vineyards, the wine label from High Plains grape growers the Seaton and Furgeson families, and Majek Vineyard & Winery. Slate Mill's winemaker, Josh Fritsche, also produces his own label wine under the name Tatum Cellars, a brand he launched in 2012.

The Slate Mill Wine Collective property features a historic homestead, barn, and smokehouse restored to its original 1800s condition, which is commemorated with a State Historical Marker. It also boasts a 35-acre (14.2-hectare) estate vineyard. With a second winery in development on Highway 290 and an adjacent 200-acre (81-hectare) events facility in Fredericksburg, Slate Mill Wine Collective will soon be managing 150 acres (60.7 hectares) of estate vineyards planted with 35 grape varieties. These Hill Country grapes will be used for Slate Mill's wine production, but also for clients and partners who will have the option of selecting from these very varieties to use in their winemaking endeavors.

Southold Farm + Cellar

Fredericksburg
southoldfarmandcellar.com

Few newcomers to the Texas wine industry have made as grand an entrance as Southold Farm + Cellar. It had a national cult following even before breaking ground on its Hill Country property near Fredericksburg in 2017. That's because the brand, which is owned by Regan and Carey Meador, existed long before it made its way to Texas. In 2012, Carey Meador launched Southold Farm + Cellar winery on Long Island, New York. The native Texan had made his career in advertising in New York City before pivoting to winemaking on the North Fork of the island. What began as a passionate taste for wine while living in New York evolved into a curiosity for making it. As Long Island had already developed a reputable wine industry, Meador and his wife, a Long Island native, left the Big Apple. Meador took a job as an apprentice for a local producer before developing a plan to launch his own brand.

Within a short time, Southold Farm + Cellar garnered critical acclaim for its low-intervention, fresh, and juicy wines. As a loyal following developed, Meador began to see his wines on restaurant menus throughout the east and west coasts. But a few years into the endeavor, the Meadors were faced with zoning conflicts that prevented them from expanding. Seeing no alternative, the two decided to move their family (including their two young children) and their whole operation to a 62-acre (25.1-hectare) property in Fredericksburg.

Meador works closely with Robert Clay Vineyards, an organically farmed site in the Mason County area in the sandstone soils of the west side of the Hill Country. He also purchases fruit from growers in the High Plains. In addition, he has planted a 16-acre (6.5-hectare) estate vineyard, working first with rootstocks to see which are best suited to his rocky, limestone-rich hillside site. In the coming years, he has plans to plant a vineyard at his family's ranch near Sonora in far west Texas.

Meador continues to use a minimal intervention approach with his wines, working to retain the acidity of the grapes to reveal the vibrancy of the fruit. In Texas, that means harvesting earlier than most other producers. This also keeps alcohol levels from creeping too high in the wines, making for more refreshing offerings to enjoy in the often oppressive Texas heat. He employs wild yeast fermentation and uses neutral vessels such as concrete, stainless steel, and neutral oak to let the fruit express its natural character. The result is a selection of wines that are infinitely drinkable and approachable.

These wines resemble the light, fruity wines of the Loire Valley and Beaujolais. Try the Foregone Conclusion, an herbaceous, red-fruited Alicante Bouschet, or the Weapon of Choice, a lighter, yet sophisticated style of Cabernet Sauvignon.

Spicewood Vineyards

Spicewood

spicewoodvineyards.com

A long-lived producer, boasting two separate lives, Spicewood Vineyards was founded in 1990 by Ed and Madeleine Manigold on an idyllic plot of land about 37 miles (59.5 kilometers) west of Austin in Spicewood. The property is located in an ancient river bed on the edge of the Llano Uplift with an individual fault running through the area, creating a displacement of more than 400 feet (122 meters). The unique topography has resulted in a complexity of soils, ranging from sand to red, loamy clay on top of limestone. Between 1992 and 1996, the Manigolds planted the majority of their 17.5-acre (7-hectare) estate vineyard at an elevation of about 900 feet (274 meters) on a gentle slope that receives strong, southeasterly breezes. This early planting included the standard Bordeaux red and white varieties, along with Zinfandel and Chardonnay. In 1995, the Manigolds officially opened their winery with a newly constructed production facility. From the beginning, the Manigolds were committed to producing 100 percent estate wines. But over the following decade, the couple began to wind down their efforts with the winery, seeking someone new to take it over.

In 2007, Ron Yates, a young law school graduate, had convinced himself that making wine was the best career path. After all, he had grown up watching his cousin, Ed Auler of Fall Creek Vineyards, pave the way for wine production in the Texas Hill Country. Yates spent time with the Manigolds, and over several conversations, he convinced the winery owners that he had the dedication and ambition to be a suitable successor to their winery. In partnership with his parents, Ron and Glena Yates, and sister, Kara Dudley, Yates took over the winery and has grown the brand while remaining as close as possible to the Manigolds' tradition of estate-grown wines—though not exclusively.

Today, Ron Yates serves as the overall chief of "doing everything", from self-distributing his wines to local restaurants and retailers to pruning vineyards, hosting events, and working alongside his winemaker, Todd Crowell.

The vineyards have expanded to include a total of 32 acres (13 hectares) under vine with nine of the original 17 acres (6.9 hectares) still in production. About 60 percent of Spicewood's wines are from the estate vineyard, while the remaining grapes come from the High Plains and west Texas. Tempranillo remains a consistent star of the estate program, as does Sauvignon Blanc, which fares unusually well in the vineyard's microclimate. The Good Guy red blend of Tempranillo, Graciano, Merlot, and Cabernet Sauvignon is a crowd favorite, and in a good vintage, varietal Syrah is exceptional.

Stone House Vineyard

Spicewood
stonehousevineyard.com

Since 2006, this boutique winery about an hour northwest of Austin has specialized in one particular red wine made entirely from the American hybrid grape Norton. Owner Angela Downer Moench, who comes from the Barossa Valley in Australia, settled in the Texas Hill Country nearly two decades ago. While Stone House does produce wines with grapes from Napa Valley as well as the Barossa Valley, where Moench also continues to maintain vineyard sites, her Texas wine made from Norton, called Claros, has become a hallmark of the industry. The wine is a consistent crowd-pleaser that is dark and concentrated with notes of dried plum and fig along with clove, spice, and savory herb, black tar, and leather. In addition to trying the Claros, the Norton-based Port-style wine, Scheming Beagle, is well worth a taste.

Wedding Oak Winery

San Saba
weddingoakwinery.com

Located in the northernmost stretches of the Hill Country, Wedding Oak Winery was founded in 2012 by Mike McHenry, an established grape grower in the region, who partnered with a few entrepreneur friends and investors who wanted to become an integral part of the emerging Texas wine scene. Named for a historic 400-year-old tree called the Wedding Oak Tree, which grows about 2.5 miles (4 kilometers) northwest of the small town of San Saba, Wedding Oak's original location is a 1926 building adjacent to the 10,000-case winery.

102   THE WINES OF SOUTHWEST U.S.A.

Committed to producing 100 percent Texas-grown wine, Wedding Oak sources fruit from both the High Plains and the Hill Country, including from McHenry's own four-acre Cherokee Creek Vineyard. With a winemaking team that includes Seth Urbanek as winemaker and noted Texas wine expert Penny Adams, a former advisor for the Texas AgriLife Extension, as viticulturist, Wedding Oak is steadily offering a portfolio of quality wines. Try the Texedo Red, a rustic blend of Dolcetto and Touriga Nacional with red fruit, pipe tobacco, and spice from the High Plains, or the elegant Sangiovese with notes of macerated cherries and leather.

William Chris Vineyards
Hye
www.williamchriswines.com

Although William Chris Vineyards (WCV) only arrived on the scene about a decade ago, the speed at which this notable producer has blazed a trail and pushed the Texas wine industry forward is nothing short of admirable. Like many unlikely business gambles, William Chris, a partnership between two stalwart Texans, is the result of a far-fetched idea shared over a few beers, or in this case, a few vodkas. A young Chris Brundrett had spent his first few years after gaining his degree at Texas A&M University kicking the tires of a few different career ideas that wouldn't keep him tied to a desk. Having worked at a winery in the Texas Hill Country for a short time, he began to feel the itch to develop a brand of his own. But he didn't want to jump into the deep waters of running a winery on his own; he wanted to do it with someone with know-how. Enter William "Bill" Blackmon, one of the foremost experts on the differences between growing grapes in the High Plains and the Hill Country. Blackmon had also been on the hunt for a business partner, someone who shared the goal of offering wine drinkers an authentic flavor of Texas.

A High Plains native, Blackmon grew up as part of a multi-generation farming family. Witnessing the ups and downs of row-crop farming from his parents, he quickly learned that making a living based on the whims of Mother Nature was not for the faint of heart. Having experienced several detrimental weather events with cotton farming by the time he had entered college at Texas Tech University, Blackmon began hearing about the potential for grapes in Texas. Armed with this knowledge, his family opted to plant 13 acres (5.3 hectares) of Chenin Blanc and Cabernet Sauvignon on their property. The venture proved relatively successful for the better part of a decade—not that it didn't meet with relative hardship, such as frosts, hail, and wind. But in 1991, a devastating frost hit the area and killed everything in the family vineyard, reducing it to the ground. The experience was such a blow to Blackmon that he almost walked away from the occupation altogether.

Instead, he moved to the Hill Country. Although this region contends with more substantial risk from disease pressure, Blackmon confesses that farming in the High Plains—with the driving winds and the gritty sand blowing around—had taken it out of him. In 1996, he was given an opportunity to work with Becker Vineyards, and he took it. In the coming years, he would also manage various other vineyards and consult on wine projects. But in 2009, he was approached by Chris Brundrett to start something of his own.

Blackmon's extensive knowledge coupled with his humility and his experience of just how gut-wrenching the lows of the profession can be is exactly what Brundrett was looking for in a partner. As it stands, he is the ever-optimistic frontman of the partnership, while Blackmon is the pragmatic, behind-the-scenes farmer. As it would turn out for WCV, the partnership was a match made in heaven. Starting small, WCV began with a production of just a few hundred cases. It sourced fruit from a few local vineyards as well as High Plains fruit from private plantings such as the Hunter Vineyards, owned by Blackmon's longtime family friends and home to perhaps the best and most surprising Merlot in all of Texas.

With virtually no marketing or advertising, word of WCV spread fast and far, bringing curious wine enthusiasts from all over the state to the casual tasting room in a small, early-twentieth-century house. And the wine spoke for itself. Today, WCV has expanded beyond just tasting room sales to include a robust wine club business, statewide retail, distribution run from the premises, and numerous side projects such as Yes We Can, a canned wine partnership with the owners of Lost Draw Cellars.

William Chris Vineyards sources 100 percent Texas fruit, not only from the Hill Country and the High Plains, but also from numerous growers in all corners of the state, east, west, and further south, to sustain its growing wine program. In the cellar, a team of winemakers led by Tony Offill and overseen by Brundrett and Blackmon has taken advantage of both Old-World and progressive winemaking techniques, and works with a wide range of coopers for a strategic barrel program.

104   THE WINES OF SOUTHWEST U.S.A.

Brundrett and Blackmon see themselves as winegrowers with a dream of sharing with people the true taste of Texas. Wines are made with low intervention and often showcase single vineyard expressiveness, regional identity, or varietal distinction. Among the varietals that best represent WCV's efforts are the Mourvèdre, Sangiovese, Tannat, and Merlot. (Barrel-fermented Roussanne and various releases of *pétillant naturel* are also well worth seeking out.)

TEXAS HIGH PLAINS AVA

Burklee Hill

Levelland

burkleehillvineyards.com

Burklee Hill is one of the many examples where a High Plains grape-growing family has complemented its farming operation by making wine. Owners Chace and Elizabeth Hill have been growing grapes since 2002 and later joined with two other winemaking families to form Trilogy Cellars. In 2017, the couple struck out to produce wine under their own label, and the result has been an overwhelming success. Burklee Hill wines are clean and varietally expressive, balancing approachability with complexity. Try the earthy Montepulciano with notes of tobacco and rich red and dark fruit, or the Courtney, a lovely aromatic blend of Viognier and Marsanne that offers a vibrant backbone through to the finish.

English Newsom Cellars

Lubbock

englishnewsom.com

English Newsom Cellars is a winery that has taken on many names, owners, and iterations before becoming the successful business it is today. Founded as Teysha Cellars in 1987, the winery soon fell into bankruptcy, but was resurrected as CapRock Winery, named after the region's famed geologic formation. Celebrated winemaker Kim McPherson made CapRock wines for more than a decade before moving on to launch his own brand. Although CapRock would see ownership change hands again in the decade that followed, in 2013 it experienced a glimmer of new hope when Tommy and Jana English stepped in. Through their efforts to reinvigorate the business, the winery solidified some of its missing pieces, hiring winemaker David Mueller, formerly of McPherson Cellars, and partnering with longtime High Plains grape grower Steve Newsom, of Newsom Family Farms. The new partnership resulted in a new name and a new story.

The majority of the fruit for Newsom English Cellars is sourced from Newsom's vineyards; supplementary grapes come from other local High Plains growers. In 2020 winery production had risen to 10,000 cases. The winery uses the remainder of the 80,000-case capacity to help other producers as a custom crush facility. Try the crisp, beautifully balanced Roussanne or the earthy and approachable Tempranillo.

Llano Estacado Winery
Lubbock
www.llanowine.com

The importance of this winery cannot be overstated. It is the first commercial Texas winery established in the state's modern wine era. It is also one of the largest, with an overall production of about 150,000 cases (although Ste. Genevieve technically produces significantly more in total cases, little of it is actually made from Texas fruit). In addition, it remains a relevant and forward-thinking leader in the industry.

Llano Estacado was born out of the experimental work of curious Texas Tech professors Bob Reed and Dr. Clint "Doc" McPherson, two gentlemen who are arguably the fathers of Texas wine as we know it today. As part of a chemistry class project, the two began making wine, the results of which were encouraging enough to prompt the two professors to launch a winery—with the help of a few investors. In 1976, Llano Estacado was established, with its first releases of 1,300 cases hitting the market in 1977. Over the next four decades, Llano Estacado would increase in not only production, but also recognition as a purveyor of quality Texas wine.

Today, the winery does produce a significant amount of non-Texas wine, particularly for its value-level, high-production retail wines. But the 40 percent of Llano Estacado's production that is from Texas is among the best in the state. Although the winery has a small vineyard adorning the outskirts of its tasting room—more for show than for overall production—the majority of the grapes sourced come from vineyards in the High Plains as well as from Dell Valley Vineyards in far west Texas. Winemaker Jason Centanni has leveraged the unique high desert soils of this particular vineyard to reveal its unique terroir. This is best represented in the Tempranillo, Chardonnay, and Syrah.

106 THE WINES OF SOUTHWEST U.S.A.

Centanni posits that his overall goal is to make varietally correct, consumer-friendly wines. While this may be the case, the wines go beyond being consumer friendly; there is certainly a greater depth and complexity to be found in many of his wines. This is what makes Llano Estacado such an essential part of the Texas wine industry.

Llano Estacado works with a wide range of grape varieties, with a focus on warm-climate cultivars such as Tempranillo, Sangiovese, Syrah, Mourvèdre, Viognier, and Marsanne, which have proved to be workhorse varieties for the winery. Cabernet Sauvignon and Petit Verdot have shown more promise than many might have expected. There have also been great successes with cooler-climate cultivars such as Chenin Blanc, Riesling, and Sauvignon Blanc.

Texas grape growers: wine is farming

Unlike many wine regions, where the vineyards are usually located in the same place as the wineries, the majority of Texas wine grapes are grown in the Panhandle. Conversely, the overwhelming majority of wineries are elsewhere in the state, with the primary concentration in the Hill Country. This means the industry relies on two types of producers. The first is an increasing number of farmers, many of whom come from a long line of cotton-farmers, who have planted a growing number of vineyards to supply winemakers (the second key producer) with quality grapes for making wine. Though there is a large percentage of vineyard planting in the Hill Country and within other pockets of the state including Tyler, Washington County, and in west Texas, the vast majority of Texas grapes are in the High Plains, with around 4,000 acres (1,619 hectares) planted within this one region. Vineyard acreage in the Hill Country area is estimated to be around 1,000 (405 hectares).

Below is a snapshot of some of the most notable High Plains and Hill Country vineyards, from whom many of the state's top wineries source their grapes.

Texas High Plains AVA

Alta Loma Vineyard
Brownfield, Terry County
Owned by: Ronnie Floyd and Bobby
   Jo Floyd, Ronny & Gale Burran
Established: 2016
Acres: 10 (4.04 hectares)

Bingham Family Vineyards
Meadow, Terry County
Owned by: Cliff and Betty Bingham
Established: 2004
Acres: 175 (70.8 hectares)

Blackwater Draw Vineyards
Brownfield, Terry County
Owned by: Summit Hogue
Established: 2014
Acres: 25 (10.1 hectares)

Diamante Doble Vineyard
Tokio, Terry County
Owned by: Jet Wilmeth
Established: 2000
Acres: 175 (70.8 hectares)

Krick Hill Vineyards
Levelland, Hockley County
Owned by: Chace and Elizabeth Hill
Established: 2002
Acres: 52 (21 hectares)

La Pradera Vineyards
Levelland, Hockley County
Owned by: Michael and Barbara
   Paddack, Andy and Lauren
   Timmons
Established: 2010
Acres: 100 (40.5 hectares)

Lost Draw Vineyards
Brownfield, Terry County and
   Lubbock County
Owned by: Andy Timmons
Established: 2006
Acres: 68.5 (27.7 hectares)

Newsom Vineyards
Plains, Yoakum County
Owned by: Neal and Janice Newsom
Established: 1986
Acres: 150 (60.7 hectares)

Canada Family Vineyards
Plains, Yoakum County
Owned by: Dwayne and Brenda
   Canada
Established: 2005
Acres: 20.5 (8.3 hectares)

Farmhouse Vineyards
Meadow, Terry County
Owned by: Nicholas and Katy Jane
   Seaton, Anthony and Traci Furgeson
Established: 2010
Acres: 112 (45.3 hectares)

Lahey Vineyards
Brownfield, Terry County
Owned by: Matt Adams
Established: 2014
Acres: 1,000 acres (404.7 hectares)

Lepard Vineyards
Brownfield, Terry County
Owned by: Russell Lepard
Established: 2007
Acres: 74 (30 hectares)

Narra Vineyards
Brownfield, Terry County
Owned by: Nikhila, Koteswari, and
   Nagarjun Narra
Established: 2014
Acres: 130 (52.6 hectares)

One Elm Vineyard
Tahoka, Lynn County
Owned by: Brad Stewart
Established: 2014
Acres: 5 (2 hectares)

108 THE WINES OF SOUTHWEST U.S.A.

Phillips Vineyard
Brownfield, Terry County
Owned by: Tony and Madonna
  Phillips
Established: 2015
Acres: 30 (12.1 hectares)

Reddy Vineyards
Brownfield, Terry County
Owned by: Vijay, Subada, and Akhil
  Reddy
Established: 1997
Acres: 270 (109.3 hectares)

Soleado Vineyards
Seagraves, Yoakum County
Owned by: the Nelson family
Established: 2015
Acres: 10 (4 hectares)

Texas Hill Country AVA

Enchanted Rock Vineyard
Fredericksburg, Gillespie County
Owned by: John and Susan Curtis
Acres: 3 (1.2 hectares)

Parr Vineyards
Grit, Mason County
Owned by: Robb and Dilek Parr
Established: 2006
Acres: 20 (8.1 hectares)

Robert Clay Vineyards
Mason, Mason County
Owned by: Dan and Jeanie
  McGlaughlin
Established: 1996
Acres: 20 (8.1 hectares)

Salt Lick Vineyards
Driftwood, Hays County
Owned by: Scott Roberts
Established: 2006
Acres: 60 (24.3 hectares)

Tallent Vineyards
Camp Air, Mason County
Owned by: Drew Tallent
Established: 2000
Acres: 60+ (24.3 hectares)

Tio Pancho Vineyard (Bear Vineyards)
Bend, San Saba
Owned by: Noelle Montag
Established: 1996
Acres: 9 (3.6 hectares)

McPherson Cellars

Lubbock
www.mcphersoncellars.com

The son of Texas wine pioneer Doc McPherson, Kim McPherson has deftly slipped into his father's shoes as one of the patriarchs of the modern Texas wine industry. The Lubbock native received his under-graduate degree in food science before moving on to the University of California, Davis to study enology and viticulture. He worked for a few years at Trefethen Winery before returning to Lubbock in 1979 to be a winemaker for Llano Estacado Winery. During his time there, he made the first Texas wine to win a double gold medal at the San Francisco Fair. McPherson left Llano Estacado to pursue his own interests for a few years before joining CapRock as head winemaker in the late 1990s. He would stay here, making wine and consulting for up-and-coming producers before launching his brand, the eponymous McPherson Cellars label, in 1998. At the time, he used the CapRock facilities to build his brand, but severed ties as the producer continued to struggle, opening his winery and tasting room in 2008 out of the historic Coca Cola bottling plant in downtown Lubbock.

Today, McPherson Cellars produces about 15,000 cases and is one of the most widely recognized Texas producers in the country, with a substantial presence on retail shelves and restaurant menus throughout the state and distribution beyond Texas borders into seven states. Much of McPherson's success is due to his business strategy of bringing wine to the general public at an approachable price (most of his wines retail for between $13 and $18). In contrast, the majority of the wines found throughout Hill Country tasting rooms are well above $30. To him, getting his wines into the hands of more consumers is what will ulti-mately win the day.

But McPherson has done more than make good wine at a reasonable price. He has also served as a mentor to any and every new winemaker seeking advice and assistance in the past decade. It's a sizable list. In fact, ask any quality Texas wine producer who helped them get their start, and you'll hear the name Kim McPherson. This generosity, combined with his business savvy, has been one of the driving forces behind tak-ing the quality of Texas wine to the next level. It helps that those who he has mentored are doing their part to push the message forward and carry the weight as well.

McPherson was one of the first in the state to champion warm-climate varieties, including those from Southern France, Italy, and Spain. Among his choice varieties, he sees the most promise in Sangiovese, Carignan, Cinsault, Mourvèdre, Roussanne, Viognier, and Picpoul Blanc. Since the beginning, he has championed regionality for the High Plains, man-aging his father's Sagmor Vineyards (13.5 acres—5.46 hectares—of which 10 acres, or 4.05 hectares, are in production) as well as working alongside the myriad growers he buys his fruit from to improve farming techniques.

McPherson's wines evoke an Old-World feel, using less oak influence, striving for moderate alcohol levels, not over-extracting in the cellar for color or alcohol, and managing tannin structure through farming. Try the Les Copains Red, a delightful little Rhône red blend of Cinsault, Mourvèdre, Carignan, Syrah, and Grenache fermented in stainless steel and aged in oak for a few months, the single variety Cinsault, or the crisp, elegantly structured Chenin Blanc.

Reddy Vineyards
Brownfield
reddyvineyards.com

Since the late 1990s, viticulturist and soil scientist Dr. Vijay Reddy has become one of the largest grape growers in the Texas High Plains. A fifth-generation farmer from southern India, Reddy came to the U.S. in 1971 and obtained a doctorate in soil and plant science. Since then, he has run a soil consulting business along with farming cotton and peanuts. But in 1997, he was encouraged to plant grapes. Today, he manages more than 400 acres (161.9 hectares) with 32 varieties. In 2019, he and his family launched their own wine brand. As evidenced by the winery's first vintages, the new wine program, led by winemaker Lood Kotze, is revealing quality offerings showcasing refined blends and varietal selections exclusively from the estate vineyard. Try the Field Blend, a proprietary blend of the vineyard's French and Italian red varieties, or the TNT Red Blend, a brawny blend of Touriga Nacional and Tempranillo.

WEST TEXAS

Brennan Vineyards
Comanche
brennanvineyards.com

Located midway between the Hill Country and the High Plains in Comanche County, Brennan Vineyards has garnered a reputation as one of the state's top-quality producers. Founders Dr. Pat and Trellise Brennan of Fort Worth originally purchased the property as a second home. It is located at one of the oldest remaining homesteads in Texas, dating back to 1876. On it is the McCrary House, which was built by James Madison McCrary, who owned the local general store and cotton gin and served in the Minutemen Rangers and later the Frontier Battalion. James McCrary and his wife, Ella, went on to raise nine children in the Austin Stone home alongside Indian Creek, which still stands today and has been registered as a Texas Historical Marker.

In 2002, the Brennans decided to plant Viognier, Syrah, and Cabernet Sauvignon and planned to sell their grapes to their good friend Dr. Richard Becker, of Becker Vineyards. But as the Brennans dug deeper into the process of growing grapes, the notion to begin their own brand took hold. In 2004, they began construction on a winery and in 2005 opened a tasting room in the historic McCrary House. At first, they sought advice and support from friends in the industry, such as Dr. Becker and Kim McPherson, but in 2007 they brought in winemaker Todd Webster, who had completed studies in viticulture and enology at Texas Tech University and Washington State University.

About half the fruit for Brennan Vineyards wines comes from their two Comanche County vineyards. The Newburg Vineyard includes 21 acres (8.5 hectares) of Viognier, Semillon, Muscat of Alexandria, Mourvèdre, Tempranillo, Cabernet Sauvignon, Ruby Cabernet, and Nero D'Avola. Their 10-acre (4-hectare) Comanche Vineyard includes Viognier, Syrah, Malbec, and Nero D'Avola. The remainder is sourced from vineyards in the High Plains. Taking what each vintage gives him, Webster does his best to make wines expressive of place and balanced with elegance and finesse. Worth trying is the Reserve Viognier for its intense stone fruit characters wrapped in chalky minerality, the Mourvèdre Dry Rosé, the earthy and robust Nero D'Avola, and the luscious Tempranillo Reserve.

Rancho Loma Vineyards (RLV)
Coleman
rlv.wine

While most people are familiar with the big Texas cities of Austin, Houston, and Dallas, along with a few popular outposts such as Marfa and the Hill Country, few have heard of Coleman. This tiny town, boasting a population of about 4,500, sits three hours northwest of Austin near Abilene and Buffalo Gap. It is here, on their ranch in the middle of nowhere, that two former commercial filmmakers, Robert and Laurie Williamson, managed to create a destination restaurant in

2003. Along the way, they helped to revitalize the culture of a small west Texas town. By 2012, the two had added a boutique inn, and in 2015 launched a casual pizzeria in the heart of Coleman. By this time, Rancho Loma had drawn national media attention, and began luring travelers from all over the country.

This idyllic western outpost served as the inspiration for the launch of a Texas wine brand reflective of the modern Texas authenticity Rancho Loma had brought to life. In 2015, Ed and Roberta Brandecker, frequent visitors to the restaurant, partnered with the Williamsons to bring this idea to fruition. The Brandeckers enlisted the help of Dr. Ed Hellman, professor of Viticulture and Enology at Texas Tech University, to help develop the estate vineyard.

Other current owners of the venture include Tony Bowden, Dr. Tom Headstream, and Tom Munson. By 2016, they had launched a portfolio of promising wines and opened a sleek, modern tasting room in downtown Coleman.

In recent years, the Williamsons have relinquished their role in the business, allowing the Brandeckers to build on the brand's established quality. In early 2020, RLV opened a second tasting room in Fort Worth, and all wine production has been moved to Lubbock. One hundred percent of the fruit for RLV is sourced from the High Plains.

Under the guidance of French wine consultant Guenhael Kessler, RLV has shown itself to be a serious contender in terms of quality and consistency in the Texas wine industry. Try the summery Très Rosé, for its bright Provençal style, or the Toro Tempranillo, a full-bodied red with notes of cola and blackberry.

Val Verde Winery
Del Rio
valverdewinery.com

Claiming the distinction of being the oldest continuously running winery in the state, Val Verde Winery is one of the industry's historic treasures. Val Verde is located in Del Rio, on the Texas–Mexico border. When it was established, in 1883, the area had just become a part of the United States. The winery was born after Italian immigrant Frank Qualia, who had migrated from Castellanza, near Milan, in northern Italy, discovered native Black Spanish grapes flourishing throughout the region of far southwest Texas. Del Rio wasn't exactly in his plans for settlement. He had hoped to join a few of his countrymen in Mexico City as part of a new colonization program. But in the 1880s, Mexico found itself suffering political unrest, and working conditions for Italian immigrants became less than ideal.

At the time, Qualia was only 18 years old, and having decided to abandon the plan to work on the colony, he was forced to rely on whatever resources he could find. While many of the men in his party opted to work on the expanding railroad or move further north to find work in San Antonio, Qualia discovered the township of San Felipe Del Rio (present-day Del Rio). It was located at the point where the future railroad tracks were to meet the Rio Grande. The local settlement here had already begun to expand, with farms in cultivation and ample water from the San Felipe Springs.

Young Frank was given a generous opportunity to lease land from a local landowner, where he decided to capitalize on his family's long tradition of winemaking and establish a vineyard and winery. Many in the region had already planted vineyards with Black Spanish (Lenoir) grapes, and Qualia did the same, naming the winery Val Verde after the county in which it is located. One reason for Val Verde's unique longevity is that it managed to survive the 16 long years of Prohibition, a feat achieved by selling grapes and obtaining a permit to make wine for sacramental purposes.

In 1935, Qualia's son, Louis, brought in cuttings of Herbemont, an American hybrid. The juice from the reddish-brown grape is a rusty, amber color so the Qualias used it to make a white wine for a few years. The vines turned out to be susceptible to chlorosis, but the family was able to save some by grafting them onto Mustang rootstock.

Frank Qualia passed away in 1936, leaving the operation to his son, Louis, who, in turn, passed the winery on to his youngest son, Thomas ("Tommy"), in 1973. To complete the generational lineage, Tommy's son Michael is set to take over operations in a few years. Today, the Qualia vineyard still has a few Herbemont vines as well as some Blanc Du Bois but is almost exclusively Black Spanish, including vines that date back more than 120 years. The winery also makes wines using grapes from vineyards in the High Plains and Escondido Valley.

Perhaps the winery's most prized offering is the Don Luis Tawny Port, which shows off the richness of Black Spanish with inky, concentrated dark-fruit flavors.

114 THE WINES OF SOUTHWEST U.S.A.

SOUTH TEXAS

Haak Vineyards & Winery

Santa Fe

www.haakwine.com

A maverick of Texas wine in more ways than one, Haak Vineyards & Winery has played a vital role in the industry, proving that both Blanc Du Bois and Black Spanish can make beautiful, well-crafted wines. Owners Raymond and Gladys Haak took a leap of faith more than 20 years ago when they staked a claim with their 3-acre (1.2-hectare) Blanc Du Bois vineyard located in Galveston County, only 20 miles (32 kilometers) from the Gulf Coast. Since then, Haak has excelled in consistently delivering beautifully balanced Blanc Du Bois—in both dry and off-dry styles. He and his winemaker, Tiffany Farrell, also make red wines such as Malbec, Cabernet Sauvignon, and Tempranillo sourced from Reddy Vineyards in the High Plains. Even more compelling are the dessert wines. Since 2006, Haak has made a name for himself with two different Madeira-style wines. These wines are produced in a special *estufa* oven, constructed by Haak himself. Following a brief time in barrel after fermentation, the barrels are brought to age at 115°F (46°C) in the *estufa* for three months. This high-temperature aging process imparts elegant, nutty, caramel flavors to the Blanc Du Bois, and a decisive fig and orange marmalade character to the Jacquez. Haak is the only Texas winery, and one of only a few in the entire country, that may legally produce, label, and sell a fortified Madeira wine. Try the Haak Blanc Du Bois (dry) with floral aromatics and notes of grapefruit and apricot, and the Blanc Du Bois or Jacquez Madeira, both of which are among the state's most celebrated dessert wines.

Messina Hof Wine Cellars

Bryan

messinahof.com

The most important thing about Messina Hof Winery, is not its longevity in the Texas industry or the fact that it is one of the state's largest producers of Texas-grown wine or even the countless medals won over the years—although all of these things are true. Perhaps what is actually most meaningful about Messina Hof, and the two people who brought it to life, Paul V. and Merrill Bonarrigo, is that family heritage, resilience, and gracious hospitality are at the core of everything this longstanding winery embodies. And perhaps that's really all that needs to be said.

Coming from a very long line of Sicilian winemakers, Bonarrigo was raised in the Bronx, watching and learning as his immigrant grandparents made wine in their apartment for family consumption. His grandfather had brought his family over from the seaside town of Messina in 1927, establishing new roots in New York.

According to Bonarrigo, for more generations than he can count, the first son in his family is named Paul. In his early career as a physical therapist, he was persuaded to set up his practice in Bryan, Texas, the home of Texas A&M University. There he met his wife, Merrill, whose German family heritage can be traced back to the Bavarian town of Hof. Early on, the two purchased a 100-acre (40.4-hectare) plot of land and named the homestead after where their families were from—hence Messina Hof. In 1977, they began helping Ron Perry, a young student at Texas A&M University, with an experimental vineyard project that he was doing for his thesis on grape feasibility in Texas. (That study was replicated in other parts of the state and helped establish research and results that carried the Texas wine industry forward during the 1970s.) Inspired by the project as well by his family's history of making wine, Bonarrigo continued the vineyard past the experimental stage and began to play with the idea of winemaking. After a few vintages of personal winemaking, the Bonnarigos began to collaborate with grape growers in the High Plains, and decided to launch a commercial winery. It is one of the pioneering wineries of the modern Texas wine era along with Llano Estacado Winery in the High Plains and Fall Creek Vineyards in the Hill Country. Today, the 20-acre (8-hectare) estate vineyard primarily includes Black Spanish (Lenoir), which goes into the winery's 'Port' production program. The majority of their grapes come from the High Plains as well as from vineyards within 50 miles (80.5 kilometers) of the Messina Hof Estate Winery in Bryan. When available, they also source from vineyards in the Hill Country and Texoma.

The original estate winery in Bryan now includes a tasting room, restaurant, event space, gift shop, and bed and breakfast. The Bonarrigos have also expanded to locations in Fredericksburg and Grapevine and have broken ground on a property near Houston. The Fredericksburg outpost includes the Manor Haüs Bed and Breakfast, as well as a tasting room; Grapevine's urban winery, housed in the historic Wallis Hotel, offers nine wines on tap.

116   THE WINES OF SOUTHWEST U.S.A.

From the very beginning, the Bonarrigos have radiated a magnetic sense of warmth and welcome not only at their winery locations but also through their involvement in industry leadership and retail and restaurant outreach. They have fostered a loyal staff who carry on the mission of hospitality that has seen countless marriage proposals, weddings, special events, and anniversaries celebrated at their properties by customers over the years. But perhaps this element of embracing wine lovers one-and-all is best translated through their wide range of wine offerings.

Messina Hof works with 32 grape varieties in Texas, which are used for blends and varietal releases in a variety of styles. Early on, Merrill even devised a color-coded key for the wine labels to indicate the sweetness level. While many producers try to manage a narrower range of wines, Messina Hof has taken the approach of catering to all taste preferences. To them, the diversity of their offering means they reach more consumers who might be new to Texas wine, giving them a positive experience that will persuade them to continue to explore.

Carrying on the family tradition, the Bonarrigos handed ownership to their son, Paul Mitchell Bonarrigo, and his wife, Karen, in 2012. While the founders still participate in promotional engagements for the winery, the next generation of owners has brought new life and ambition to the brand.

The winery's annual production is around 92,500 cases, including 60,000 wines bottled under the Messina Hof label, with remaining wine allocated to a bulk wine program used to support and help develop other Texas wineries. While Messina Hof may be known for a wide range of offerings, favorites include Sagrantino, which has proved to be an exceptional red grape in both the High Plains and at the estate vineyard, in chemistry and wine quality. Messina Hof has also earned an unofficial reputation as the "Riesling house of Texas"—Riesling manages to do well in the High Plains—offering three styles, including a Dry Riesling, the off-dry Father and Son Riesling, and the Late Harvest Angel Riesling. In addition to these, try the Private Reserve Cabernet Franc, a luscious offering with a soft palate and notes of mocha and dark, cordialed fruit, or the GSM, a consistent crowd pleaser that offers bright, red fruit characters with earthy undertones.

EAST TEXAS

Enoch's Stomp
Harleton
enochsstomp.com

Founded in 2004 by Altus Koegelenberg and Jon Kral, this east Texas outpost is part of a 90-acre (36.4-hectare) property sprawled among the rolling hills of Harleton (about an hour east of Tyler, near the Louisiana border). Koegelenberg, a fifth-generation grape grower from South Africa, moved his family to East Texas and turned his attention to growing grapes in the Piney Woods of Texas. He was fortunate to meet Jon Kral, originally from the Pacific Northwest, who had a degree in chemistry and had also taken an interest in winemaking. In 2005, the business partners planted 12 acres (4.8 hectares) of vineyard on the estate in Harleton, opting for American hybrids, including Lenoir, Blanc Du Bois, Cynthiana, Norton, Champanel, and Villard Blanc. Although the wines are made with grapes that are less widely known to the general public, over the past decade the winery has encountered a steady flow of visitors eager to discover something new. Try the flagship Blanc Du Bois, offered in both a dry and an off-dry style. Go for the dry, which is balanced and approachable, offering soft notes of stone fruit and flowers. Also worth seeking out is the lush Cabernet Sauvignon.

Fairhaven
Hawkins
fairhavenvineyards.com

Out near Hawkins—about two hours east of Dallas—since its establishment in 2004 Fairhaven Vineyards has made a name for itself by specializing in French–American and American hybrid grapes. Owner R.L. Winters, a horticulturist, has spent much time working to carry on the research of T.V. Munson. The vineyard is now home to 11 acres (4.4 hectares) growing around 25 grape varieties, including Chambourcin, Norton, Lomanto, Favorite, Lenoir, Blanc Du Bois, Brilliant, Wine King, Delicatessen, Nitodal, and a few *Vitis vinifera* selections such as Albariño, Tempranillo, and Pinot Blanc.

Much of Winters' focus has been on developing the "Heritage Series" of super-hybrid wine grapes; heat and drought tolerant, disease-resistant varieties. As part of his work, he has become a purveyor of hybrid vines, the majority of which he sells to vineyards in the American South and

Mid-Atlantic regions. The majority of Fairhaven wines are estate grown, though a small percentage of fruit is sourced from other parts of Texas and from Napa Valley. The 2009 Lomanto was the first wine from an American hybrid grape to win an international award since 1873. A particular star from the Fairhaven portfolio is the Chambourcin, which is consistently jammy and juicy, with bright acidity and notes of tart dark fruit on the palate.

Kiepersol Estates Winery

Tyler
kiepersol.com

Although it is located far to the east of where most of the Texas wine action is, Kiepersol has played a key role in the Texas wine story. Founded by Pierre de Wet in 1998, it became the thirty-fourth bonded winery in the state of Texas. Originally from South Africa, de Wet grew up in a multi-generational farming family. When he came to east Texas in the 1980s, he continued with farming, eventually securing the opportunity to develop a 67-acre (27.1-hectare) vineyard, winery, distillery, restaurant, bed and breakfast, and RV park.

When he established the vineyard in 1998, he named the venture Kiepersol Estates Winery, after the area of South Africa's Eastern Transvaal where his father's farm was located. In 2000, the Kiepersol tasting room opened, and in the following two decades, the project would expand into the multifunctional operation it is today. Although consultants from the Texas Agriculture Extension Service warned de Wet against planting *Vitis vinifera*, arguing that the disease pressure in east Texas made the region better suited to French–American hybrids, he dismissed the advice. Instead, he planted Cabernet Sauvignon, Merlot, Syrah, and South Africa's star white variety, Chenin Blanc. In later years, he would add Alicante Bouschet, Cabernet Franc, Malbec, Mourvèdre, Muscat di Alexandria, Pinot Gris, Ruby Cabernet, Sangiovese, Sauvignon Blanc, Semillon, Tempranillo, Viognier, and Zinfandel. De Wet's determination paid off. Kiepersol Estates has become particularly well known for its bold and brawny Stainless Cabernet, fermented and aged, as the name suggests, in stainless steel tanks. The Chenin Blanc, which is labeled as Steen, is a sentimental reference to de Wet's homeland.

In 2016, de Wet unexpectedly passed away, leaving a legacy of hard work and determination. Today, Kiepersol Estates Winery is owned and operated by his two daughters, Marnelle de Wet Durrett and Velmay de Wet Power, their families, and a team of loyal staff that includes winemaker Adam Lee.

Los Pinos Ranch Vineyards
Pittsburg
www.lospinosranchvineyards.com

A collaboration project between four east Texas families, Los Pinos Ranch opened in 2002. The property is a 40-acre (16.2-hectare) estate, of which 17 acres (6.9 hectares) are under vine. The vineyards feature Blanc Du Bois and Black Spanish, two varieties that thrive in the east Texas soils, tolerating the humidity with little disease pressure. With an annual production of about 15,000 cases, Los Pinos Ranch also sources fruit from High Plains growers with a focus on Italian varieties, including Sangiovese, Vermentino, and Dolcetto. Winemaker Amulfo Perez uses a light hand in his production process, employing native yeast fermentation for most of the portfolio. Try the aromatic Blanco Grande made from estate Blanc Du Bois, the dry rosé-style Rosato or the fresh, easy-drinking Sangiovese Riserva.

8

WHERE TO EAT AND STAY IN TEXAS

TEXAS HILL COUNTRY AVA

The Texas Hill Country has received wide acclaim as one of the state's most popular tourist destinations. Scattered among the winding roads between the region's charming towns are myriad shops, restaurants, bed and breakfasts, and recreational activities that draw visitors from all over the country. Not surprisingly, they come for the wine, too.

Stay

Camp Comfort
Comfort
camp-comfort.com
Located in the charming German township of Comfort, this collection of warm, renovated cabins high above Cypress Creek offers cozy respite and friendly service.

Camp Lucy
Dripping Springs
camplucy.com
Directly west of Austin, Dripping Springs is often referred to as the Gateway to the Hill Country. As you meander a few winding backroads to Camp Lucy, the basic stresses of life seem to melt away. Though initially an event space for weddings and special occasions, this rustic-chic ranch now boasts private, modern farmhouse cabins for overnight visitors to linger in the idyllic atmosphere.

Cotton Gin Village

Fredericksburg
cottonginlodging.com

Adjacent to the Cabernet Grill, the quaint Cotton Gin Village offers seven nineteenth-century log cabins, along with seven new-construction cottages with massive cedar wood four-poster king beds, wood-burning fireplaces, and a small kitchen. One of the best parts of the stay is the breakfast picnic basket delivered to the door each morning with warm, house-made baked goods, quiches, and yogurt with fruit.

Walden Retreats

Johnson City
waldenretreats.com

This 96-acre (38.8-hectare) outpost offers Hill Country glamping at its finest. The expansive private property near Johnson City features private cabins that overlook the winding bends of the Pedernales River. The luxurious safari-style "tents" are chic and cozy—an ideal escape during the fall.

Dine

Cabernet Grill

Fredericksburg
cabernetgrill.com

As one of only two restaurants in the state of Texas to boast an all-Texas wine list, the Cabernet Grill offers an enticing menu of sophisticated fare with bold flavors inspired by the heritage of the Hill Country. The jalapeño-stuffed grilled quail and the braised beef short rib are consistent crowd favorites.

Dai Due

Austin
daidue.com

One of the pioneers of farm-to-table dining in Austin, Dai Due is the vision of celebrated Texas chef Jesse Griffiths. Central to his cooking approach is the philosophy that every ingredient must be sourced locally and in season. By extension, that philosophy also applies to the restaurant's beer and wine list. As such, Dai Due is the second of only two restaurants in the state with an all-Texas wine list. Dining here is genuine proof of the adage "what grows together goes together".

High's Cafe and Store

Comfort
highscafeandstore.com

A great stop for breakfast or lunch, not only is the cafe and its adjoining shop bright and charming, but the low-key, farm-fresh fare is addictively delicious. Flavorful salads and house-made soups are always a good option as are the peppered pimiento cheese and the crab cakes. Don't leave without a decadent, chocolatey macaroon, or a tart and tangy lemon bar.

Otto's

Fredericksburg
ottosfbg.com

With a distinctive menu of German-inspired Hill Country fare, Otto's is a pleasant locale with friendly service and a well-appointed international wine menu.

The Parlour

Johnson City
theparlourjctx.com

A casual wine and cheese bar from the owners of Hill Country winery Southold Farm + Cellar. Choose from a selection of Texas and non-Texas wines to go along with a diversity of cheese selected by expert cheesemonger and general manager Courtney Schwamb. An assortment of seasonal snack offerings is available throughout the week.

Pecan Street Brewery

Johnson City
pecanstreetbrewing.com

A casual brewpub featuring a wide selection of house-made beers and a well-curated Texas wine menu. Pizzas and burgers are always on point, but be sure to start with an order of creamy beer cheese served with crispy tortilla chips.

Salt Lick BBQ

Driftwood
saltlickbbq.com

This Texas icon has been serving up southern-style barbecue for more than four decades and has become a sought-after destination for locals and out-of-town visitors alike. Its legendary circular stone-wall open pit

124 THE WINES OF SOUTHWEST U.S.A.

is always ablaze, with juicy selections of brisket, ribs, and sausage perpetually on display. Also worth noting, the adjoining Salt Lick Cellars offers a wide range of Texas wine and beers.

Tillie's
Dripping Springs
tilliesdrippingsprings.com
Located about 20 miles (32 kilometers) west of Austin, Tillie's restaurant is housed in a century-old building that was once a town hall in Vietnam. The flavorful, innovative approach to traditional American cuisine uses local, seasonal ingredients, frequently fusing them with Asian, Mexican, and classic European flavors.

Vaudeville
Fredericksburg
vaudeville-living.com
This home-design showroom and shop has the added benefit of a basement bistro offering a seasonal chef-driven menu for both lunch and dinner.

The San Antonio missions: remembering the Alamo, and Texas Independence at San Jacinto

The history of wine in the Southwest owes its origin to Spanish conquest from Mexico north, into what is now New Mexico and Texas, throughout the seventeenth and eighteenth centuries. In 1682, missionaries fleeing the revolts of New Mexico Puebloans arrived in El Paso, constructing missions along the river. In Texas, in particular, what remain of the Spanish missions established throughout the state are considered historical treasures. At first, the missions were established as part of a political play of New Spain as it expanded into the frontier. Initially, they were just bases for further exploration. But with the arrival of the French in the 1680s, the Spanish crown felt pressure to establish its presence. It launched a campaign to begin building missions and presidios, military forts to defend the missions, further into the region that is now Texas. At first, this included far into east Texas, but with persistent resistance from the French and from indigenous tribes, the Spanish issued a renewed focus along the Rio Grande, the Gulf Coast, and at San Antonio.

At the Presidio de San Antonio de Béjar (present-day San Antonio) in particular, the Spanish established a series of missions along the San Antonio

River. (Less successful attempts also occurred in central, southeast, and west Texas.) Today, five of the original missions around San Antonio still stand as reminders of the Spanish presence in Texas history. In 1978, four of the missions were designated the San Antonio Missions National Historical Park: Mission Concepción, Mission San José, Mission San Juan, and Mission Espada. The fifth (and most famous) mission in San Antonio, the Alamo, is not part of the Park. It is located upstream from Mission Concepción, in downtown San Antonio, and is owned by the State of Texas.

The Alamo is most famed for the dramatic twelve-day skirmish that began there on February 23, 1836. The Spanish army, led by Mexican General Antonio López de Santa Anna, attacked a regiment of the Texas army. The battle was hard-fought and ultimately won by Santa Anna in a bloody battle on March 6. More than 200 Texans are said to have perished, with only one survivor. But while it is the Alamo that has remained etched in the minds of most Texans—particularly the battle cry, "Remember the Alamo"—the battle represents loss and humiliation for the Texas military. It was the Battle of San Jacinto, which took place on April 21, 1836, near present-day Houston, which ultimately won Texas its independence from Mexico. Texas historians consider the Battle of San Jacinto to be one of the most decisive battles in American history. It led to Texas annexation into the United States in 1845 and to the Mexican-American War, which led to the U.S. acquisition of New Mexico, Arizona, Nevada, California, Utah, and parts of Colorado, Wyoming, Kansas, and Oklahoma. This accounts for almost one-third of the area of the present-day United States.

On July 5, 2015, the San Antonio Missions National Historical Park, along with the Alamo Mission, was designated a UNESCO World Heritage Site. Visitors can walk throughout the missions, admiring their historic architecture and stepping back into a pivotal time in the history of Texas.

See

Enchanted Rock State Natural Area
Fredericksburg
tpwd.texas.gov/state-parks/enchanted-rock

Perhaps one of the most visited Hill Country recreational spots, for locals and visitors alike, Enchanted Rock is a geologic wonder that is best seen in person. This massive pink-granite dome, which climbs to 425 feet (129 meters), is a natural playground for hiking, rock climbing, and camping.

Luckenbach

Fredericksburg
luckenbachtexas.com

No visit to the Hill Country is complete without a stop at Luckenbach. Music lovers in particular will want to visit this storied pitstop, which earned international fame as the central locale for the celebrated Waylon Jennings and Willie Nelson song, "Luckenbach, Texas Back to the Basics of Love". Today, it still serves as a cherished music venue to which aspiring singer-songwriters flock as a sort of religious pilgrimage. The dance hall presents more than 300 events a year featuring Americana, honky-tonk, western swing, rockabilly, roots rock, alt-country, and blues. Located off Farm to Market Road 1376, the few old buildings, including a general store and a dance hall, represent the whole of Luckenbach. It may not seem like much, but there's no better way to experience the soul of Texas than on a weekend night of country dancing under the party lights strung around the dance hall.

Willow City Loop

Ideal for a Sunday drive, or a rigorous ride for those who enjoy cycling the rolling hills, this Hill Country roadway near Fredericksburg offers some of the best local scenery. Not far from Enchanted Rock State Natural Area, the Willow City Loop affords some of the best views of the granite outcrops left from the Llano Uplift. A visit during March, in early spring, presents the best opportunity to see fields of bluebonnets, the Texas state flower, in full bloom. Take 16N (Llano Street) out of Fredericksburg and proceed 13.3 miles (21 kilometers). Turn right on RR1323 and continue 2.79 miles (4.5 kilometers) to Willow City before turning left onto Willow City Loop. This is a private ranch road, so be sure to respect private property. The 13-mile (21-kilometer) loop winds around to reconnect with Highway 16.

TEXAS HIGH PLAINS AVA

Located more than 400 miles (644 kilometers) north of the Hill Country, the Texas High Plains is a less popular destination for tourism than its regional vinous cousin, but as more and more wineries begin to open here, that will no doubt begin to change.

Stay

The Overton Hotel
Lubbock
overtonhotel.com

Conveniently located near Texas Tech University, the Overton is one of Lubbock's most deluxe accommodations. Adorned with modern Texas décor, the 15-story landmark hotel offers a glimpse of High Plains heritage with friendly hospitality. Rooms are spacious and comfortable, offering large windows that afford ideal views of the west Texas sunset.

Dine

Cocina de la Sirena
Lubbock
lasirenacocina.com

This Latin-inspired locale offers a wide range of Spanish and Latin-American dishes with bold flavors and finesse. Start with crisp-fried "poblano fries" with a sweet chile vinaigrette for dipping, and then try the stacked enchiladas for the main course. To sample a diversity of things, opting for a few *platos pequeños* (small plates) to share among friends is the way to go. While the wine menu is well-appointed, the boozy cocktails are expertly executed.

Evie Mae's Pit Barbecue
Wolfforth
eviemaesbbq.com

One of the state's top barbecue joints, offering peppery, mouthwatering brisket along with house-made green chile sausage, ribs, and all the "fixin's" (coleslaw, potato salad, and pinto beans, of course). Don't leave without having banana pudding for dessert.

La Diosa Cellars
Lubbock
ladiosacellars.com

Just across the street from McPherson Cellars (see p. 108) in downtown Lubbock, Kim McPherson's wife, Sylvia, has owned and operated this Spanish-inspired gem since 2004. The eclectic atmosphere and

Old-World vibe have made La Diosa one of Lubbock's most treasured local hang-outs. Classic tapas and authentic Spanish fare are all expertly executed and served in a vibrantly colorful dining room adorned with equally vivid works of art.

Above: Summer showers threaten vineyards during harvest at D.H. Lescombes Vineyards in the Mimbres Valley AVA of New Mexico.

Below: Lush vineyards thriving at La Viña Winery in the Mesilla Valley AVA of New Mexico.

Above: Forbidden Desert Vineyards is located in Engle, New Mexico, just between the Mesilla and Middle Rio Grande Valley AVAs of New Mexico. Here, Noisy Water Winery's Jasper Riddle grows Pinot Noir, Chardonnay, Chenin Blanc, and Cabernet Sauvignon.

Below: The vineyards of Vivác Winery rest beneath the sandstone Barrancos Blancos mountain range just north of Santa Fe, New Mexico.

The wind machines of Lost Draw Vineyards in the High Plains of Texas. Late spring and early fall freezes are a perpetual challenge throughout the Southwest making wind machines a useful tool for grape growers.

Above: The Sagmor Vineyard in Lubbock, Texas is home to 34-year-old Cabernet Sauvignon and Sangiovese vines planted by Clint "Doc" McPherson, one of the fathers of the Texas wine industry. Today, these vines are farmed by McPherson's son, Kim, of McPherson Cellars.

Below: The Kuhlken Vineyard, a family vineyard initially planted in the 1990s in the Texas Hill Country, which led to the inception of Pedernales Cellars.

Above: The Willcox AVA lies in the southwest corner of Arizona and is home to more than 70 percent of the state's vineyards.

Below: The Southwest Wine Center at Yavapai College in Arizona's Verde Valley offers hands-on opportunities for students to learn not only winemaking and viticulture, but also the business of running a winery.

Above: The sun sets over vineyards in Arizona's Sonoita AVA.

Below: Tight clusters of Cabernet Sauvignon ripen in the hot Arizona sun at Callaghan Vineyards in the Sonoita AVA.

Above: The Colorado River cuts through Colorado's Grand Valley AVA.

Below: Mount Lamborn rises high over the West Elks AVA of Colorado.

The Book Cliff Mountains are a defining feature of Colorado's Grand Valley AVA.

PART 3
ARIZONA—DESERT WINE

9

HISTORY

The Arizona wine industry reportedly grew by 1,940 percent between the years of 2003 and 2017, a number which seems incredibly implausible at first glance. But a more in-depth look into the evolution of Arizona wine in the past ten years alone reveals a surge in education, vineyard plantings, quality production, and investment in the future. While Arizona wine may remain in the vinous shadows today, the shroud is beginning to lift. With a committed focus on quality and philosophy of collaboration among the industry's most progressive producers, the potential seen in Arizona wine means there is a strong possibility that it will soon emerge as the shining star of the Southwest.

EARLY ARIZONA WINE

As in the rest of the Southwest, early Arizona wine is tied to the Spanish missions of the sixteenth century. However, whereas evidence of seventeenth-century vineyard plantings in New Mexico, Texas, and California is fairly concrete, little has been found regarding any plantings in Arizona.

Early Arizona vineyards have been documented as far back as 1703 when a Jesuit missionary, Eusebio Francisco Kino, planted grapes in the Tucson area. One of Arizona's more comprehensive historical accounts comes from historian Erik Berg's 2018 *Journal of Arizona History* article, "Equal Age for Age: The Growth, Death, and Rebirth of an Arizona Wine Industry, 1700–2000". This resource looks deeply into the history of its early wine years. This, along with Thomas Pinney's two volumes, *A History of Wine in America: From the Beginnings to Prohibition*,

132 THE WINES OF SOUTHWEST U.S.A.

and *A History of Wine in America: From Prohibition to the Present*, help solidify Arizona wine's history.

Following the Mexican–American War, which came to an end with the 1848 Treaty of Hidalgo, Arizona originally joined the United States as part of the Territory of New Mexico. The geographical boundaries of the state remained in limbo for more than a decade. The southern half first seceded from the Union as the Territory of Arizona, part of the Confederate States, in January of 1862. Just over a year later, in February of 1863, the Federal government of the United States reclaimed the territory, with boundaries that would form the basis of its current borders.

During this time, silver and gold prospectors had made their way to the Santa Rita Mountains, establishing the Santa Rita mine and the Heintzelman mine in the Cerro Colorado Mountains in the southern parts of the state in 1856. In the years that followed came several other mines throughout the territories. The mining boom is significant for the influx of settlers that came with it, most of whom had a big thirst for beer, spirits, and wine.

Not surprisingly, it is not long after Arizona's admittance to the U.S. and its uptick in mining operations that mention of a significant vineyard planting near present-day Phoenix surfaces. According to Berg's report, in 1867 an irrigation company, Swilling Irrigating and Canal Company, formed under the ownership of J.W. "Jack" Swilling in 1867. As part of a growing community built up around the Salt River, a significant amount of farmland had been cultivated. By 1872, wine grapes had been documented around the town of Mesa, about 20 miles (32 kilometers) east of Phoenix. Among these early grape growers were Daniel and Samuel Bagley, who eventually opened the area's first commercial winery in the early 1880s. Within a decade, the Bagleys had been joined by other producers, west of Phoenix in the Salt River Valley, which was recorded as having produced more than 50,000 gallons (189,270 liters) of wine and brandy per year.

Cochise and the Chiricahua Apache

Among the many indigenous tribes that flourished throughout the Southwest in centuries past, the Apache groups are known to have put up the fiercest, and most long-lasting resistance to the invasion of their homelands by Spanish and other European settlers.

Traditionally, the Chiricahua (pronounced CHEER-uh-KAH-wa) Apache faction consisted of three to five bands, which were divided into local groups of

10 to 30 extended families, each occupying individual territories. Their liveli-hood depended on hunting wild game and gathering wild plant foods, which meant they frequently moved on in order to avoid exhausting their food sources. Although no central organized tribe existed, each band was led by a chief and remained a socially cohesive group.

The Spanish settlers gave them the name "Chiricahua", which means ei-ther "chatterer" or "grinder". Other names given to the tribe have included Mimbreños, Coppermine, Warm Springs, Mogollon, Pinery, and Cochise Apache. The bands referred to themselves as Nde, Ne, Néndé, Héndé or Hen-de, meaning "the people, men".

They roamed the rugged, mountainous regions west of the Rio Grande into southeastern Arizona and northern Mexico. By the late sixteenth century, the Spanish were building settlements and missions throughout the Southwest. While at first the Chiricahua were willing to trade with them, the Chiricahua soon became skeptical when the Spanish enslaved many on the missions and pushed for Christian conversion. There followed centuries of battles be-tween the Chiricahua and the Spanish, and later the Mexicans, who became independent of Spain in 1821. Despite many efforts by the Spanish to subdue them, the Chiricahua held on to their sacred homeland.

In 1848 Mexico was defeated in its war with the United States, surrendering a large section of the Southwest, which included much of the Apache terri-tory in Arizona and New Mexico. With the westward expansion of the United States in the late nineteenth century, the Apache now had another contender vying for their land. From the 1850s until 1875, many Apache groups were forcibly moved to several reservations in Arizona and New Mexico, except for the Chiricahua, who continued to resist. When the United States attempted to move the settled tribes to a reservation further from their native land, the initi-ative was met with great resistance. Before long, fighting and raiding resumed.

In the Chiricahua Mountains, the Chiricahua Apache took refuge under the leadership of Chief Cochise (pronounced coh-CHEES). Here, the formidable warrior sought refuge with about 1,000 followers among the rocky maze of the mountain formation. Posting lookout scouts from various points on the towering pinnacles of rock, Cochise could spot enemies in the valley below and sweep down without warning, making the Chiricahua stronghold a peril-ous journey for anyone who passed.

Historical accounts document Cochise perpetually warring with Mexican troops and western frontiersmen who relentlessly encroached upon Apache homelands. In 1861, after years of struggle and loss, despite the warrior's fervent resistance, Cochise met with the U.S. military under the white flag of truce to deny charges that one of his people had abducted a white child. Despite the flag of truce, the commanding U.S. officer ordered the six attending chiefs seized and bound because they would not confess. One was killed, and four were caught as they resisted arrest. Cochise cut through the side of the tent and fled, suffering multiple bullet wounds during his escape. His revenge was swift and bloody. Within weeks, all frontier encampments were raided and their inhabitants massacred, sending the clear message that the Apache were at war with the white man. The Apache Wars that followed lasted for more than three decades. U.S. military troops eventually abandoned the area and were reassigned to take part in the American Civil War.

Convinced of victory, and emboldened by the aggressive attack strategy, Cochise allied with Mangas Coloradas, leader of the Mimbres Apache of New Mexico, to later defend Apache Pass in southeast Arizona against the Californians in 1862. On the settlers' side, the Battle of Apache Pass was led by General James Carleton, who was charged with reopening communication between the Pacific Coast and the East and removing Confederate Army soldiers from Tucson. The Californians were armed with howitzers, which quickly overwhelmed the Apache, who fled to their many refuges in the neighboring mountains. Still the Apache refused to surrender. When United States troops returned to resume the occupancy of the country after the close of the Civil War, an all-out war of extermination was carried out against the Apache.

In 1863 Mangas Coloradas was captured, tortured, and killed. Cochise and his small group of followers held on for another ten years. In 1872 he surrendered, living out the remaining two years of his life on the new Chiricahua Reservation. Upon his death in 1874, he was secretly buried somewhere among the rocks of the Dragoon Mountains, now called the Cochise Stronghold. The exact location of his burial site has never been determined.

After Cochise's death, the Chiricahua lost their reservation and were moved to the San Carlos Reservation in Arizona. Unhappy with the area, with the Apache already living there, and with their treatment by U.S. government agents, some Chiricahua escaped, leading to another decade of fighting.

In the early to mid-1800s, there were an estimated 2,500 to 3,000 Chiricahua Apache. In 1886 there were just over five hundred. By 1959 there were about 91 full-blooded Chiricahua at the Mescalero Reservation in New Mexico, but in 1990 only seventeen Chiricahua were living in New Mexico. In the 1990 U.S. Census, 739 people identified themselves as Chiricahua. In 2000 that number had increased to 1,155.

In the late 1880s, in the verdant hills of the Verde Valley, German immigrant Henry Schuerman planted a sizable vineyard along Oak Creek, not far from present-day Sedona. His winery is said to have supplied the growing mining camp of Jerome. While the trend in wine production continued to gain steam at the turn of the twentieth century, it was not long-lived. Just as other states within the Southwest were grappling with the rapid growth of the temperance movement, Arizona was also swayed. In 1915, the state enacted its own legislation banning the production and sale of alcohol. Wineries shut down, vineyards were uprooted, and according to Berg's research, Schuerman was briefly jailed for trying to sell off his remaining inventory.

THE BEGINNINGS OF THE MODERN ERA

It would be several decades, following the end of Prohibition in 1933, before Arizona's wine industry would reignite. In common with its neighboring states, Arizona's current wine industry began in the 1970s. A young soil scientist at the University of Arizona, Dr. Gordon Dutt, started to experiment with growing grapes near Oracle (about 35 miles, or 56 kilometers) north of Tucson. The vineyard was planted at an elevation of about 5,000 feet (1,524 meters) with red, ferruginous soils. Half expecting the red grapes to scorch in the radiant Arizona sun and lack any concentration of color, Dutt was happily surprised to find that he could produce wines with not only concentrated color but also vibrance and structure.

Dutt's early findings with his vineyard prompted him to join horticulturist Eugene Mielke and lead a team of researchers to study the wine-growing potential of the Southwest as part of the Four Corners Regional Commission project. By 1980, the team had released a report that strongly encouraged the planting of vineyards throughout the region. Among the team's findings within Arizona were the high potential for winter kill for vines planted at high elevations, the challenge of inadequate water at specific sites, and the presence of cotton-root rot. They were able to narrow down that the valleys and plains of southeastern Arizona, where elevations ranged from 2,500 to 5,000 feet (762 to 1,524 meters), such as in Cochise and Santa Cruz counties, were more than suitable for grape growing. The report went on to detail the climate, soil compositions, and best growing techniques, as well as the

136   THE WINES OF SOUTHWEST U.S.A.

suggested grapes that were, at the time, considered to be commercially viable for the region including Malvasia Bianca, Emerald Riesling, French Colombard, Barbera, Petite Sirah, and Cabernet Sauvignon.

In 1979, prompted by his findings, Dutt launched the state's first modern commercial wine vineyard, Viña Sonoita. The property was leased from the Babacomari Ranch, which was one of the test sites Dutt had planted during the Four Corners Regional Commission study. He likened the soils of the site to those of Burgundy's Côte D'Or, featuring weathered clay and traces of calcium carbonate. Sadly, all of these original plantings fell victim to Pierce's disease in the early 1990s and had to be replaced.[4]

At the time, Arizona laws were riddled with restrictions on wine production and sales. Wineries could only sell to distributors and were banned from direct sales to other businesses or the public. To counteract these measures, the 1982 Arizona Farm Winery Act was passed, allowing wineries to sell directly to consumers. The legislation ushered in the first licensed wineries in Arizona, including Dutt's Sonoita Vineyards.[5]

With the passage of the Arizona Farm Winery Act, a flurry of new wineries and vineyard plantings were established in the Sonoita area. To cement the nascent industry's commitment to moving forward, in 1984 Arizona became the first state in the Southwest to establish an American Viticulture Area (AVA), with Sonoita.

A few years before Sonoita received its AVA designation, Robert Webb, whose R.W. Webb Winery was located in Tucson, started looking to plant his own Arizona vineyard. He soon settled on a 40-acre (16.2-hectare) plot of land on the slope of the Dos Cabezas Mountains near Willcox, about 83 miles (133.5 kilometers) east of Tucson. In 1984, he planted 20 acres (8.1 hectares) of grapes including Cabernet Sauvignon, Merlot, and Petite Sirah. By 1986, he had doubled his production and opened a new tasting room.

"I learned that to be successful as an Arizona winery, you had to do more than make good wine," Webb wrote in an article for *AZ Lifestyle* magazine in 2010. "You had to make good wine out of Arizona fruit and market, market, market." Indeed, Webb pushed this business philosophy through for the following decade, which saw his label make

---

4   Most of the vineyards in Arizona are now free of this pesky plant pathogen, but the threat remains a persistent challenge for growers to mitigate.

5   Robert Webb, who opened the R.W. Webb Winery in 1980, claimed the distinction of opening the first winery in Arizona since Prohibition.

its way to the retail shelves of every supermarket chain in the state by 1997.

In 1989, Webb sold his vineyard to Al Buhl, a native of Staten Island, New York, whose first exposure to viticulture was in the Finger Lakes region during his undergraduate years at Syracuse University. Inspired by the work Dutt had started in the 1970s, Buhl immersed himself in viticultural research, reading extensively and seeking advice and mentorship from others in the industry, including New Mexico's Paolo D'Andrea of Luna Rossa Winery. In the following decades, he became one of the most influential grape growers in the state, mainly due to his interest in warmer climate grape varieties from Italy, Spain, and southern France as opposed to the Bordeaux varieties that many people had planted at the time. He added Viognier, Sangiovese, Pinot Gris, and Malvasia Bianca to the vineyard plantings.

For many years, Buhl sold his fruit to wineries, including Dr. Dutt at Sonoita Vineyards, and Kent Callaghan of Callaghan Vineyards, who had also been making wines from his Buena Suerte Vineyard in Sonoita since 1991. But in the early 1990s, it became clear that making wine could also be a worthwhile endeavor, especially when left with excess fruit from bountiful harvests in certain years. In 1995, Buhl formed Dos Cabezas WineWorks in partnership with Callaghan, as well as a handful of other partners. Callaghan served as head winemaker for a while before Frank DiChristofano came on board. Later, a young Todd Bostock joined the winery as assistant winemaker and would eventually end up taking over the operation.[6]

By this time, Willcox was well on its way to establishing itself as a unique wine-growing region in Arizona. As rumor spread of its relatively affordable land and its ideal growing conditions, the area began to attract more attention. In 1999, noted filmmaker Sam Pillsbury purchased 40 acres (16.2 hectares) next to Dos Cabezas vineyard, in partnership with Buhl, and planted 20 acres (8.1 hectares) exclusively to Rhône Valley varieties, including Grenache, Syrah, and Mourvèdre. In addition, famed Oregon winemaker Dick Erath of the eponymous Erath Winery took an interest in Willcox, having retired to Tucson to escape the Oregon winters. In 2004, he purchased 40 acres (16.2 hectares) of land for planting the Cimarron Vineyard in the Kansas

---

6 Buhl passed away in 2014. His vineyard is now called the Al Buhl Memorial Vineyard and is owned and managed by Caduceus Cellars. Bostock took ownership of the Dos Cabezas WineWorks winery in 2006, relocating it to Sonoita.

138   THE WINES OF SOUTHWEST U.S.A.

Settlement of Cochise County and in 2005 began planting Italian varieties such as Sangiovese, Montepulciano, and Primitivo. Today, the vineyard is owned by Todd Bostock of Dos Cabezas WineWorks.

In 2016, Willcox finally received its designation as an AVA even though the majority of vineyard plantings had already far outpaced that of Sonoita or anywhere else in the state. Today, it remains the region where most of Arizona's grapes are grown.

VERDE VALLEY

While the southern part of Arizona was fleshing out its own wine identity, further north, in the Verde Valley near Sedona, another region was also developing. Though evidence of early wine production from wild grapes dates back to the mining era of the late nineteenth century, experimentation with plantings and wine production in this part of the state started in the late 1970s. In 1977 New Yorker William Staltari moved his family to Arizona and planted grapes in Camp Verde. By 1981 he had launched San Dominique Winery, the second post-Prohibition winery in the state. Staltari would go on to play a critical role in shaping Arizona's modern wine industry, serving on the board of the Arizona Wine Growers Association, and passing the Domestic Farm Winery Bill.

Further north, in the Oak Creek area, Jon Marcus started Echo Canyon Vineyard & Winery, planting 32 acres (12.9 hectares) in 1993 and releasing his first vintage in 2001. In 2001 Oak Creek Vineyard and Winery started up, and in 2002, Maynard James Keenan began planting his first vineyards on different sites throughout the region. What followed was a flurry of producers, from Javelina Leap Vineyard & Winery in 2005 and the Dancing Apache (D.A.) Ranch vineyard planting in 2006 to the founding of Page Springs Cellars in 2004 and Arizona Stronghold Cellars in 2007, which would quickly become one of the state's largest brands. Tasting rooms have helped revitalize Jerome and Cottonwood's historic downtown areas with new restaurants and accommodations, and inspired Yavapai Community College to launch the state's first enology program.

Navigating the lineage of Arizona's wine history seems more like untangling a spider's web, rather than following a linear timeline. But the meandering path it has forged in such an accelerated amount of time— compared to the histories of the world's iconic wine regions—is perhaps

Arizona's greatest asset. Though not perfect, the established network of its tight-knit community has managed to solidify a message of quality and identity that will no doubt propel it into a stable seating next to California, Washington, and Oregon in the near future. Rather than an industry with six degrees of separation between key players, it's more like one or two degrees. The overall cohesion that has risen from this closely woven community is likely to be Arizona wine's greatest success.

10

CLIMATE, REGIONS, AND CHALLENGES

While most people associate this sunny state with the wild west, picturing cactus-riddled desert vistas and the iconic Grand Canyon, in recent years Arizona has witnessed a new awakening with glimpses of excellent wine. To date, the state has just over 115 wineries and more than 1,200 acres (485.6 hectares) of vineyards in production. Its wine industry is reported to have an economic impact of more than $56.2 million, attracting more than 600,000 visitors to the vineyards and tasting rooms of the state's three primary wine regions.

Arizona's varied landscape is unique in its ability to experience wide temperature variations during the growing season – the diurnal range can swing by as much as 50°F (27.7°C) in ripening season. Another key growing factor for Arizona vineyards is that most of them are planted at elevations of 3,500 to 5,500 feet (1,067 to 1,676 meters) above sea level.

Of its three primary viticultural regions, two have official AVA designation. Sonoita, about 161 miles (259 kilometers) southeast of Phoenix, is a rolling landscape that is home to myriad cattle ranches. The Willcox Playa, or Willcox Bench, is home to a large dry-lake basin with the Chiricahua Mountains and Dos Cabezas Mountains to the east and the Dragoon Mountains and Cochise Stronghold to the west. The northern Verde Valley constitutes the third wine-growing area in the state. Though not an official AVA, the region is on the list for its AVA designation and is expected to be approved within the next year or two. Located about two hours north of Phoenix by car, with vineyard sites planted at

141

142 THE WINES OF SOUTHWEST U.S.A.

roughly 3,000 to 5,500 feet (914 to 1,676 meters) in elevation, the area is sprinkled with nearly two dozen producers amid a handful of small-town communities.

CLIMATE

Arizona, like New Mexico, is part of the desert Basin and Range Province, which is a physiographic region spanning the inland Southwestern part of the United States and northern Mexico. Arizona's expansive, barren landscape is scattered with drought-tolerant scrub-brush and cactus, though a surprising 27 percent of the state is forest land. In fact, the world's most massive Ponderosa pine tree is located in the Prescott National Forest. The majority of the state's topography, which encompasses a total of 113,998 square miles (295,255 kilometers), was shaped by prehistoric volcanism resulting in a uniquely diverse geography. It is renowned for its jagged, narrow mountain chains, flat valleys, deep canyons, high elevations, low-lying deserts, and dramatic red-hued rock formations. The result is an almost confounding array of soils, from rocky and sandy, to red, iron-rich loam, clay, and calcium-rich caliche. The high bicarbonate levels found in these caliche layers bring a particular minerality to the wines that offsets the high pH levels also found in the soils.

With such variations in elevation in any given part of the state, Arizona is unique among the southwest's wine regions in having a wide variety of localized climate conditions. In general, the climate can be sweltering in the summers, particularly in the south-central capital city of Phoenix, which is situated at a lower elevation of about 1,086 feet (331 meters), in the bottom of a bowl-shaped depression of the state. Here, temperatures can reach more than 106°F (41°C) in summer, while in other parts of the state temperatures are lower, though still averaging well above 90°F (32°C). Winters are generally mild, with lows reaching close to 45°F (7°C) in Phoenix and 17°F (-8°C) further north in Flagstaff. The more mountainous northern portion of the state, with its significantly higher altitudes, offers an appreciably cooler climate, with cold winters and mild summers.

While the average rainfall is nearly 13 inches (330 millimeters), the majority of the year is severely dry and arid, except for two specific rainy seasons. The first is in the winter, with cold fronts influenced by the Pacific Ocean to the west, and the second is the monsoon season, which pulls tropical moisture up from Mexico to make for significant rains during the late summer—often in the thick of grape harvest. The higher humidity combined with the scorching summer heat brings lightning, thunderstorms, wind, and torrential downpours.

Late summer rains mean growers can't afford to be *laissez-faire* in the vineyards if they want to maintain quality fruit. Instead, canopy management, proper spraying, and the management of airflow through the vineyards are critical during the growing season. As winegrowers have found in recent decades, Arizona's climate is generally conducive to yielding grapes with desired ripeness and structure—more so with certain varieties than with others. However, maintaining the balance between acidity and ripeness remains a persistent challenge throughout the state, especially when the threat of heavy rain often guides the decision to pick during the most crucial part of the season.

But rainfall later in the growing season isn't the only potential problem for wine growing. As with most of the Southwestern states, late spring and early fall frosts are a perpetual threat. Many winegrowers have invested in wind machines to help mitigate frost, but it's not always a guaranteed safeguard.

The northern and southern parts of the state feature elevation gains resulting in much cooler wine-growing regions. In the southern part of the state, in the Chiricahua Mountains near Willcox, vineyard sites are planted at elevations upward of 5,300 feet (1,615 meters). Further west, in Sonoita, sites range between 4,000 and 5,000 feet (1,219 and 1,524 meters), and in the Verde Valley to the north, some of the region's highest plantings are at around 5,000 feet (1,524 meters). Despite Arizona's searing summer heat, these higher elevation sites, among the highest in the world, benefit from the dramatic shift in evening temperatures, which help slow vine respiration, allowing the grapes to retain acidity. These conditions also aid in tannin development, resulting in wines with good aging potential despite elevated pH levels. In the vineyards, growers work to combat high pH levels with a healthy canopy-to-fruit ratio that provides ripeness without any hang time necessary. Making a pass during the mid-season for shoot-thinning can also help. When yields are managed at between two and four tons per acre (4,483–8,966 kilos per hectare), the pH levels are rarely too high. Despite these efforts, acidulation in the winery is relatively standard, particularly for red wine (although some producers are choosing to live with higher pHs, in the 3.9–4.0 range).

144   THE WINES OF SOUTHWEST U.S.A.

The Grand Canyon: the pride of Arizona

Ask anyone what landmark is most associated with the state of Arizona, and invariably the answer will be: the Grand Canyon.

Considered one of the Seven Natural Wonders of the World, the canyon spans 277 miles (445.5 kilometers) along a winding stretch of the Colorado River. Over millions of years, the vividly colored, steep-sided gorge has been cut to depths of more than a mile. Located in the northern part of the state, the canyon is the result of centuries of erosion that have exposed one of the most complete snapshots of geologic history on the planet. The major geologic exposures in the Grand Canyon range in age from the two-billion-year-old Vishnu Schist at the bottom of the Inner Gorge to the 230-million-year-old Kaibab Limestone on the Rim.

Within the inner crevices and caves of the canyon, indigenous tribes have left remnants of their dwellings throughout history. Early on, the Ancestral Puebloan people made their homes here, carving out garden sites, storage areas, and habitat dwellings. But the Puebloans were not alone. Nearly a dozen tribes have cultural links to the region, including the Yavapai-Apache Nation, the Hopi Tribe, the Navajo Nation, the Havasupai Tribe, and multiple bands of the Paiute Tribe. Today, the area is home to many archeological sites that preserve the structures and artifacts from these indigenous cultures.

Early exploration of the Grand Canyon began in the mid-sixteenth century, with excursions by García López de Cárdenas under the leadership of Francisco Vázquez de Coronado, whose Spanish army traveled north from Mexico City in search of the Seven Cities of Cíbola. After traveling for six months, Coronado's army arrived at the Hopi Mesas, east of the Grand Canyon. Guided by the Hopi, the explorers had hoped to find a rumored "great river" that could serve as a waterway to the Gulf of California. Mistrustful of the Spanish newcomers, it is said that the Hopi redirected the camp away from the river, leading them to believe that the area was an unnavigable, impenetrable wasteland. As a result, Coronado dismissed further western exploration and redirected his men east to Texas. The Grand Canyon was left unexplored for 235 years.

In the nineteenth century, explorers Joseph Christmas Ives and John Wesley Powell led separate expeditions into the area. By the turn of the century, the completion of the Santa Fe Railroad brought visitors to the canyon's South Rim. Along with them came early settlements looking to capitalize on the influx of tourists. As early as 1906, United States President Theodore Roosevelt began work to have the area protected, establishing it as a U.S. National Monument in

1908. Still, it wasn't until 1919 that it was established as a U.S. National Park by President Woodrow Wilson, following more than a decade of pushback from local miners and private landowners.

Today the Grand Canyon National Park is one of the world's most visited natural attractions, drawing in nearly five million visitors per year. The South Rim of the canyon is open year-round (weather permitting) and as well as attracting casual sightseers is used as the base for activities such as rafting, running, helicopter tours, and hiking. The North Rim is generally open mid-May to mid-October, depending on weather and snowpack. The floor of the valley is accessible on foot or by mule. While experienced hikers have been known to hike rim-to-rim in one day, this endeavor is strongly discouraged by park officials for amateur hikers, due to distance, steep and rocky trails, changes in elevation, and the danger of heat exhaustion from the much higher temperatures at the bottom. One other popular way to see the canyon is via the river itself, and multiple river-raft operators take tourists down the rushing water of the Colorado River during the warmer season.

REGIONS

Arizona is home to three specified growing regions, though only two of them, Sonoita and Willcox, have official AVA designations. The third, Verde Valley, has an application pending approval as of publication date.

Sonoita AVA

Though most people pronounce it son-oy-ta, the Spanish pronunciation, soy-no-ita, is perhaps more revealing about this idyllic landscape. Meaning "I am not little", with a tone of endearment, Sonoita is a region of undulating grasslands that blanket the rolling hills, peppered by the outstretched canopies of native oaks and embraced from every direction by the Santa Rita, Mustang, Whetstone, and Huachuca Mountains. Indeed, with a landscape reminiscent of Alentejo, Tuscany, or Paso Robles, when it comes to its potential as a winegrowing region, there's nothing little about it. Once home to vast cattle ranches, the plentiful rainfall and well-draining soils prompted several winegrowing pioneers to plant vineyards here in the 1970s. Located about an hour southeast of Tucson, in the southeastern corner of the state, Sonoita was the first AVA in Arizona, receiving its designation in 1984. Though it resembles a valley, the region is more of a basin comprising the headwaters for three distinct drainages, including the Sonoita Creek to the south, Cienega Creek to the north, and the Babocaman River to the east.

Surrounded by the popular tourist towns of Sonoita, Patagonia, and Elgin, Sonoita's geography sits at elevations between 2,500 and 5,000 feet (762–1,524 meters). It is heavily influenced by its nearby mountain ranges, with a unique mixture of iron-rich reddish loam soils. Compared to the Willcox region to the east, Sonoita's soils tend to be thinner and more shallow, forcing vines to take longer to get established. For grape growers, the early years of a vineyard are a game of patience. Once established, yields tend to be lower, but the quality of the fruit tends to reveal more structure and concentration. During the growing season, high temperatures range between 80 and 90°F (27–32°C), and average rainfall is 15–20 inches (380–500 millimeters), most of which occurs during monsoon season.

Some of the state's most revered vineyards are located in Sonoita, including Callaghan Vineyards and Sonoita Vineyards. The region is home to more than a dozen wineries and tasting rooms of note, including Callaghan Vineyards, Dos Cabezas WineWorks, Rune Wines, Deep Sky Vineyards, and the new Los Milics Winery.

Willcox AVA

East of Sonoita, near the southeastern border of the state, the arid, dusty mesa of the Willcox region is a cornerstone for Arizona wine. Though it only received its official AVA designation in 2016, the district has long been one of the most productive growing regions of the state. Currently, it accounts for more than 70 percent of Arizona's grape production. A large percentage of these grapes are planted on the "Willcox Bench", an alluvial fan that elevates the vineyards along the historical Kansas Settlement farmland. The result: excellent grape-growing conditions.

The AVA covers a total area of 526,000 acres (212,864.6 hectares) within Graham and Cochise Counties, including the town of Willcox along with Kansas Settlement, Turkey Creek, and Pearce. The area is a shallow "closed basin", separated from neighboring valleys by the Pinaleño, Dragoon, Chiricahua and Dos Cabezas mountain ranges. Most of the region's vineyards are planted in the Sulfur Springs Valley and along the bases of the mountains in the area, at between 4,000 and 5,500 feet (1,219–1,674 meters). At this elevation, vineyards in the area can experience up to a 50-degree Fahrenheit variance (27.7° Celsius variation) in diurnal temperature during the growing season. This basin is reliant on its average rainfall of 13–18 inches (330–457 millimeters), most of which comes from heavy summer monsoons, to recharge its underlying aquifer. In contrast, the area surrounding it has year-round creeks and streams. The dry, desert climate benefits the grapevines by placing stress on them during the growing season, slowing the vegetative growth and adding complexity to the grapes.

The soils of the Willcox AVA are mainly alluvial and colluvial and composed of loam made up of nearly equal parts sand, silt, and clay. These loamy soils retain enough water to hydrate the vines while allowing sufficient drainage through to the aquifer. The soils are referred to as the Tubac, Sonoita, Forrest, and Frye soil types, and are not found to a great extent in the area surrounding the AVA. Compared to Sonoita, Willcox's soils are generally more productive, though there is a high degree of variation from site to site.

Most vineyards in the Willcox AVA are harvested by machine

Willcox is home to roughly 1,000 acres (404.7 hectares) of vineyard, including the Al Buhl Memorial Vineyard, planted by Arizona wine pioneer Robert Webb in 1984, and now owned by Maynard James Keenan of Caduceus Cellars and Merkin Vineyards. Nearly two dozen wineries and tasting rooms are scattered throughout the region, including Bodega Pierce, Pillsbury Wine Company, Sand-Reckoner Winery, and Keeling-Schaeffer Vineyards.

For years, the Willcox area has been a rural agricultural area devoid of much tourism infrastructure. In fact, the primary road leading to the majority of vineyard sites wasn't even paved until a couple of years ago. Producers and visitors alike had to make their way along open dirt roads that fast became muddy pits, where vehicles without four-wheel-drive would often get stuck. (Access improvements came thanks to forward-thinking producers, including Barbara and Dan Pierce of Bodega Pierce, who worked with the county to have the road paved.) More restaurants and guesthouses have sprung up in and around Willcox with the hope that they will transform the region into a weekend destination for visitors.

Verde Valley

Though the Verde Valley is not yet an official AVA, its location in the northwest part of the state boasts high elevations and ideal growing conditions, with ample water from the Verde River and well-draining soils.

Located in the red-rocked canyon lands of Northern Arizona, about two hours north of Phoenix by car, the region is home to the towns of Sedona, Jerome, Camp Verde, Cottonwood, Clarkdale, and Cornville, the geographic heartland of Arizona. The Verde Valley is part of Yavapai County, bordered in the northeast by the red-hued rock formations of Sedona, with the craggy rise of the Black Hills mountain range to the south, and, to the west, a rolling landscape peppered with the towering pines of the Prescott National Forest.

The region is home to a range of volcanic soils at higher elevations, leading to sandy and clay loam soils throughout the valley along with alluvial deposits near Oak Creek and the Verde River. The common thread for the Verde Valley with its compatriot wine-growing regions to the south is a layer of calcium-rich caliche deep beneath the topsoil. Vineyards in the Verde Valley can be found hugging the foothills of mountain rises at elevations hovering around 5,000 feet (1,524 meters) or near the valley areas of Oak Creek and the Verde River at around 3,400 feet (1,036 meters). Compared to Sonoita and Willcox, the Verde Valley is considered a lush, verdant desert region, with an abundant water source from its two primary rivers as well as a higher percentage of rainfall, which averages about 16–18 inches (404–457 millimeters) per year—mostly during monsoon season.

Winegrowing began taking off in the Verde Valley just after the turn of the millennium, starting with just a handful of vineyards and evolving to include more than two dozen wineries and tasting rooms spread throughout its 450 square miles (1,165 square kilometers). Though an application for AVA designation is still pending, the past decade has seen an increase in wine tourism that has revived the area's small towns with new restaurants, shops, and accommodations. Its proximity to Sedona, in particular, has made it a perennial tourist favorite for wine-tasting excursions. Among the region's top tasting rooms are Chateau Tumbleweed, Merkin Vineyards Osteria, Bodega Pierce, Caduceus Cellars, Page Springs Cellars, and the Southwest Wine Center.

Mavericks and pioneers

Although the three central growing regions of the state are relatively well defined, several other growers and producers have been experimenting in other parts of the landscape. These vineyard sites can be found on the outskirts of valleys and mountain areas, such as the Chino Valley, Young, Kingman, Williams and Portal, and the Chiricahua Mountains.

150   THE WINES OF SOUTHWEST U.S.A.

Considered to be on the frontier of wine growing for Arizona, this contingent of producers has become known locally as the "mavericks and pioneers".

GRAPES

Arizona continues to experiment with grape varieties, with growers looking to site-specific climatic and soil conditions to reveal what grapes will best work. With both soil and water of a high pH, the most significant challenge has been to find varieties that can retain a decent amount of acidity in the grapes throughout the ripening process.

The most recent Arizona Vineyard Survey from the U.S. Department of Agriculture was conducted in 2013. At the time there was an estimated 950 vineyard acres (384 hectares) with the top five varieties planted including Cabernet Sauvignon, Syrah, Grenache, Zinfandel, and Merlot. But a lot has happened in the industry since 2013. For one, it is estimated that vineyard plantings are between 1,000 and 1,200 acres (405–486 hectares) with around 100 producers throughout the state. More importantly, the state's leading producers continue to plant a diversity of varieties such as Vranac and Xarel-lo to see what else works well.

In general, producers have leaned into Mediterranean varieties such as Sangiovese, Grenache, Mourvèdre, Aglianico, Syrah, Tempranillo, Petite Sirah, and Graciano for wines. However, you will still find Bordeaux varieties planted throughout the state, with Petit Verdot, Cabernet Franc, and Malbec having favorable success. Most of the state's top producers have employed the practice of creating blended wines, either through co-fermentation or post-fermentation, to help find texture and balance. The red wines of Arizona can be characterized as powerful, yet restrained, with a purity of fruit, dried herbs, and a florality evocative of the region's desert flowers.

For white grapes, Malvasia Bianca has long been a star of the region, thanks to Al Buhl's early plantings. Vermentino, Picpoul Blanc, Marsanne, and Viognier have become popular Mediterranean varieties, while Chardonnay and Pinot Gris have done well in certain sites, particularly in Willcox. Other varieties such as Petit Manseng and Clairette are showing promise. Though not particularly high in natural acidity—most wines have to be acidulated in the winery—the white wines tend to have a particular mineral component, perhaps drawn from the calcium carbonate elements in the caliche in the soils.

Cabernet Franc from Callaghan Vineyards waiting to be pressed

According to Michael Pierce of the Southwest Wine Center at Yavapai Community College, the intense sun and UV exposure also make shady canopy management a crucial part of viticultural practices in the vineyard. White grape varieties such as Viognier and Malvasia Bianca need adequate shade to avoid sunburn. Most vineyards employ vertical shoot positioned (VSP) trellising, and some have opted to modify their system by using cross arms to open up the canopy and provide shade to the fruit without compromising photosynthesis. The by-product of this modification is greater airflow. This limits potassium movement from the foliage to the fruit during ripening, often an adversely high element in the grape musts after harvest. Some producers have also experimented with applying micronized gypsum, which supplies extra calcium and displaces some of the potassium in the vines. Similarly, foliar calcium applications can achieve the same goal.

> ### *Vitis arizonica*: Arizona's native grape
>
> Though it grows in areas throughout the Southwest, *Vitis arizonica* is considered the native grape of Arizona. These vines are hardy and vigorous, and they also happen to be drought-tolerant, salt-tolerant, and resistant to Pierce's disease. In fact, in December of 2019 the University of California, Davis released five new grape varieties with this tolerance, using *Vitis arizonica* as the primary cross for them. Pierce's disease is said to cost grape growers more than $100 million a year in damage. You can find these woody vines growing wild throughout the riparian areas of the southern parts of Arizona. While it's not widely cultivated in vineyards, some producers are growing a little for experimentation while others are using the rootstock to graft in other *Vitis vinifera* varieties on their own.

CHALLENGES

As can be expected with any fledgling wine industry, Arizona faced many early challenges, ranging from plant diseases such as cotton root rot to devastating hail storms. But the common theme throughout the Southwest in terms of viticultural challenges is late spring frosts. In Arizona, the rise and fall of elevations make it easy for pockets of cold air to settle into certain vineyard sites. As in Texas and New Mexico, wind machines and burning hay bales seem to be the most common lines of defense among grape growers.

Arizona is unique in having a monsoon season, which arrives in August and can last until October, just as harvest season begins. As a result, grape growers have learned to manage vineyards with this expectation in mind. Canopies are now being trained up and open, and vineyard managers are learning to be careful not to set too much fruit.

Regardless of the region, picking a site with decent air drainage is a critical factor in avoiding winter kill and spring frost. Even a couple of feet of elevation can make the difference between a great site and an impossible one. Indeed, with each new vintage, growers are learning to manage their inputs to mitigate the results of these threats better.

Aside from the viticultural challenges, the industry has spent the past decade working to determine the best grape varieties for each region. Equally important has been the effort to foster a better public perception among consumers. As is the case with most emerging wine regions throughout the country, the hardest consumers to convince to support the industry are the ones closest to it.

Perhaps one of Arizona wine's most significant hurdles is its inability to grow beyond a specific production size. As of the date of publishing, Arizona wineries are legally prohibited from making more than 40,000 U.S. gallons (151,417 liters), or about 200,000 750-milliliter bottles, without impeding their retail licenses. The largest single producer, Arizona Stronghold, makes about 36,000 gallons (136,275 liters) annually, but it is not able to produce beyond that quantity under its existing permit. In the past few years, many of the state's more progressive producers, including Callaghan Vineyards, Dos Cabezas WineWorks, Caduceus Cellars, and Chateau Tumbleweed have led the effort to change the regulations. In February 2020, legislation was introduced to the Arizona State Legislature to reduce unnecessary regulation on Arizona wineries and allow small farming operations the opportunity to grow and compete with neighboring states such as California. The legislation has yet to pass, but the topic remains pressing among producers who are hopeful of seeing progress in the coming legislative sessions.

11

ARIZONA PRODUCERS

SONOITA AVA

Callaghan Vineyards

Elgin

callaghanvineyards.com

To say Callaghan Vineyards is the cornerstone of Arizona wine would be an understatement. Not only does owner Kent Callaghan put out some of the most compelling wines in the state, he has also been one of the most trusted mentors and advisors to others in the industry for more than 30 years.

Callaghan planted his original Buena Suerte Vineyard with his parents, Harold and Karen, in June of 1990. Harold, who was already living in Sonoita, had begun experimenting with home winemaking, and his interest in taking the hobby more seriously had taken hold. Kent Callaghan, a native of Tucson, had just graduated college and opted to join his father in the endeavor. He started by taking extension courses from UC Davis to prime his knowledge of growing grapes and making wine. The Callaghans began by planting 17 acres (6.9 hectares) with Cabernet Sauvignon, Cabernet Franc, Merlot, and a little Zinfandel. But to their misfortune, the "good luck" to which the vineyard's Spanish moniker alludes was sadly lacking, as the state experienced a significant heatwave later that summer (so severe, in fact, that it shut down the Phoenix airport for a few days). The effects of the heat devastated the new planting. But it left the Callaghans to consider an essential factor: selecting grapes that were better suited for the warm climate.

Callaghan Vineyards in the Sonoita AVA

The Callaghans quickly transitioned their plantings to include varieties such as Grenache, Syrah, Barbera, Marsanne, Mourvèdre, Malvasia Bianca, Viognier, and Roussanne. They have since added Petit Verdot, Tempranillo, Tannat, and Graciano, among others. According to Kent Callaghan, what grows well in Arizona still needs exploration, but in many ways, the sky is the limit. The vineyard site sits at 4,800 feet (1,463 meters) with a clear view of the Mustang Mountains in the distance. Today, Kent and his wife, Lisa, manage the vineyard, which now has a total of about 25 acres (10.1 hectares) planted in the heart of the Sonoita region. The soils are generally iron-rich, sandy, clay loam with heavy amounts of bicarbonate, which results in vines with less vigor, and smaller crops compared to those encountered in Willcox.

On average, the winery produces about 2,000 cases annually, but in certain vintages has reached upwards of 3,500. Though you'll find single variety releases, Callaghan has leaned into the benefits of blending his wines, using the individual varieties to work as building blocks for more complete wines with structure, balance, and food-friendliness.

But Callaghan's reputation in Arizona wine extends far beyond just winemaking. Ask just about anyone who has entered the wine industry in the southern part of the state in the past decade, and they'll likely reference him as one of the key figures who helped or gave them advice. Guided by the general philosophy that a rising tide raises all ships, his generous support, guidance, and willingness to stand among others who are working to push the quality of Arizona wine has been an integral part of shaping the identity the industry has today.

Of the varietal releases try the Graciano, Grenache, and Cabernet Franc. Among the blends, Lisa's Blend is a vibrant and aromatic blend of Viognier, Riesling, and Malvasia Bianca. Padres is a robust and spicy red blend of Graciano, Tannat, and Cabernet Sauvignon.

Deep Sky

Elgin
deepskyvineyard.com

Trying to transfer the vision of an idyllic mountain vineyard tucked against the Andes Mountains in Argentina's Uco Valley over to the southern part of Arizona may seem like an ambitious endeavor. But it's precisely the dream Phil and Kim Asmundson realized in 2011 when they planted a 20-acre (8.09-hectare) vineyard in Willcox. Following a celebratory vacation to the heralded region south of Mendoza, the family was struck by the beauty of the landscape and the depth of its wines. Rather than taking a few photos and souvenirs home with them, they opted instead to buy a parcel of a 1,500-acre (607-hectare) vineyard through The Vines Resort and Spa of Mendoza, a vineyard and wine collective with 130 individual owners from all over the world. For more than a decade, The Vines has offered a multi-layer opportunity for wine enthusiasts and would-be producers to own vineyard parcels, manage and harvest their own grapes, make wine, develop a brand and bring their wine to market. The endeavor allows a hands-on introduction to the wine business.

But dipping their toes into the world of winemaking in Argentina was only the beginning. Inspired by the landscape, soil types and elevations of Willcox, the family decided to see how well Arizona could grow Malbec—along with a few proven Rhône varieties. They named the vineyard Deep Sky, referencing the vastness of the Arizona night sky, and the endless opportunities it represents.

158 THE WINES OF SOUTHWEST U.S.A.

In 2017 they built a winery and tasting room in Elgin, enlisting the help of up-and-coming winemaker James Callahan of Rune Wines to head up the cellar team. The Asmundsons released their first vintage in 2014, a red blend of Syrah, Petite Sirah, Mourvèdre, and Counoise. Today, more than half the grapes harvested are sold to other wineries, including Rune Wines and Merkin Vineyards. The remaining harvest goes into Deep Sky's production, which is made up exclusively from estate vineyards.

Try the red-fruit dominant Willcox Malbec called Big Bang, a lighter, less concentrated offering compared to a more conventional Uco Valley Malbec, or the rich and robust Gravity, a Syrah and Petite Sirah red blend.

Dos Cabezas WineWorks
Sonoita
www.doscabezas.com

If you had asked him 20 years ago Todd Bostock never would have imagined he'd be running one of the state's longest-standing brands. Yet today, this gentle giant of Arizona wine, along with his wife, parents, and kids is doing precisely that, and pushing quality in the industry forward at the same time. A Phoenix native, Bostock managed to find his way into Arizona wine at just the right time. Having little more than a curiosity about wine, he first toyed with a home wine kit in 2001, an experience that, as he recalls, was "interesting, but not at all satisfying". His curiosity awakened, he enrolled in the extension program at UC Davis and tried getting a job at a winery in California. At the time, the coursework from Davis was given via VHS cassette tapes. Bostock remembers these videos prompting him to try an Arizona wine for the first time. It was a Dos Cabezas Sauvignon Blanc. From the very first sip, he was hooked.

Inspired by the thought of wine in his home state, he reached out to Kent Callaghan in the hope of securing an internship. Callaghan suggested he get in touch with Dos Cabezas WineWorks, which at the time was located in Willcox. Dos Cabezas has technically been around since 1995. It is the original brand launched by the late Al Buhl in partnership with Kent Callaghan, Bob Loew, Jon Marcus, and Don and Katherine Magowan. Though Callaghan served as the operation's first winemaker, the role was later passed on to Frank DiChristofano. At Callaghan's suggestion, Bostock contacted Buhl about apprenticeship opportunities, and by October of 2002, he found himself volunteering his time as an assistant winemaker. A year later, at the tender age of 25, he had worked his way up to winemaker.

In 2004, witnessing the serious commitment to wine Bostock had made, and seeing promise in his hard work, his parents, who had no previous wine experience, decided to plant the Pronghorn Vineyard in Sonoita.

In 2006, Buhl and his partners were ready to discontinue their work with Dos Cabezas, and allowed Bostock to take it over, which he managed to do with the help of his family.[7] The Bostocks moved production to Sonoita to be closer to the family vineyard. Buhl's original Dos Cabezas vineyard was sold to Maynard James Keenan and Eric Glomski, who soon after formed Arizona Stronghold. Today Keenan is sole owner of the vineyard, which has been renamed the Al Buhl Memorial Vineyard.

In total, the Bostocks farm two separate vineyards to produce their wines. In Sonoita, the Pronghorn Vineyard is 15 acres (6 hectares) planted at about 4,800 feet (1,463 meters). Pronghorn was initially planted in 2004 and is home to several different varieties, including Malvasia Bianca, Picpoul Blanc, Roussanne, Aglianico, Graciano, Petit Verdot, Petite Sirah, Tempranillo, and Vranec. Further east in Willcox, the Cimarron Vineyard was originally planted in 2005 by noted Oregon winemaker Dick Erath, who befriended Bostock and sold it to the family in 2011. Located at Kansas Settlement in Cochise County, the vineyard is 37 acres (15 hectares) sitting at about 4,300 feet (1,311 meters). In total, there are more than two dozen varieties planted, including the same varieties found in the Pronghorn Vineyards, with the addition of other Italian and French varieties, ranging from Barbera and Sangiovese to Cabernet Franc and Counoise. Additional white varieties include Albariño, Muscat, and Riesling.

Today, Dos Cabezas WineWorks produces about 5,000 cases annually, with a goal of making wines that accurately and deliciously convey the story of each vintage in southern Arizona. These wines are distinctive, yet accessible, pairing well with the food, weather, and culture of the state. Bostock has been at the forefront of experimenting with different varieties to see what works best for the different Arizona wine regions. His steady, humble persistence, hard work, and sanguine optimism have put him in the thick of moving the needle forward for Arizona wine.

---

7 Buhl and Frank DiChristofano retired, while third partner, Sam Pillsbury, went on to start Pillsbury Wine Company.

160   THE WINES OF SOUTHWEST U.S.A.

In truth, Bostock is not alone in adopting this approach, which is embodied by all involved in the business: his wife, Kelly—who serves as co-winemaker and general manager of all things behind-the-scenes—his parents, Paula and Frank, his sons, and the wine crew. Bostock would be the first to say that without them, Dos Cabezas would not be possible. This quiet leadership has positioned him as an integral figure in the story of Arizona wine.

Though you'll find a handful of single variety releases here and there in their portfolio, Dos Cabezas places emphasis on blends. Try the El Campo and El Campo Blanco, both blends which show Sonoita well, while the Aquileon red blend or the Meskeoli white blend represent the terroir of Willcox.

Los Milics Winery
Elgin

Most people who know Pavle Milic recognize him as the spirited, affable character commonly seen flitting around from table to table at his FnB restaurant in Scottsdale, like a hummingbird casing a garden in full bloom. He's spent the majority of his career in hospitality, creating the wine lists and managing front-of-house for some of the country's top restaurants, from New York to Napa Valley—and a few places in between. But his time at FnB as co-owner, general manager, and wine director for the past decade has given him a clear vision for his next endeavor: making Arizona wine. It's a dream that has been simmering in his head for nearly 16 years. Speaking of Napa Valley, he recalls a visceral reaction to first entering the region, one in which all stress seemed to melt away. It's a feeling he's been trying to recapture ever since he left California.

Upon visiting Sonoita for the first time, Milic was struck by the same feeling of calm and tranquility he had encountered in Napa. Just as you pass the U.S. Border Patrol Station on State Highway 83, the whole valley opens up before you—the Mustang Mountains to the east and the Santa Rita Mountains to the west—with undulating grasslands between, and a wide-open sky above. According to Milic, when he first encountered this place, he felt like anything was possible.

In 2018, with the help of a silent investor, he and his wife, Carla, purchased a vineyard in Elfrida, south of Willcox, which had already been planted with Tempranillo, Grenache, Graciano, and Mourvèdre. They also bought a 20-acre (8.1-hectare) Sonoita property in the foothills of the Mustang Mountains where they could plant a second vineyard and build a winery. The entrance to the property sits at about 4,900 feet (1,493 meters) and rises to 5,000 feet (1,524 meters). Here he has planted Petit Verdot, Montepulciano, Marsanne, Vermentino, Petit Manseng, Cabernet Franc, and Tannat (he also has plans for Gamay).

Knowing the world of wine and trying to produce it are two entirely different things, which is why Milic sought guidance from two of the most trusted leaders in the industry, Kent Callaghan of Callaghan Vineyards and Todd Bostock of Dos Cabezas. In 2014, he launched his own label, Los Milics, using Bostock's facility—and guidance—at Dos Cabezas WineWorks. Using fruit from his Willcox vineyard, Milic aims to make wine that best represents the region.

Construction for the Los Milics winery began in early 2020. It will include a 7,400-square-foot (687.5-square-meter) winery and a separate, sleek, state-of-the-art tasting room, restaurant, and boutique hotel, the whole of which is set to be a cornerstone in positioning the Sonoita region as one of the country's most sought-after wine destinations.

Los Milics wines combine freshness and complexity, employing neutral barrel fermentation for almost all wines as well as carbonic maceration for reds and some aging in concrete egg. Try the rich, elegantly structured Tempranillo or the earthy, red-fruited Grenache.

Rune Wines
Sonoita
www.runewines.com

A native Arizonan, James Callahan found his way into wine after fostering a fascination he had developed for the subject while working in a few fine-dining establishments after college. Before long, he found a job in Tempe for an urban winery making Cabernet from Napa. As he began to understand the winemaking process, Callahan often found himself surprised that he was making wine 10 minutes from his home in Tempe. He couldn't help but wonder about doing something like this on his own.

Deciding to seek outside opportunities, he took on work in Washington with Waters Winery, followed by a harvest season in the Wairarapa region of New Zealand, working with Pinot Noir in Martinborough, and later, as a harvest worker for Kosta Browne Winery in the Russian River Valley, where he worked his way up to cellarmaster. Despite his eye-opening adventures, Callahan couldn't shake his desire to open up something of his own.

162   THE WINES OF SOUTHWEST U.S.A.

While he could have stayed on to grow his career in California, he happened on a job opening in Willcox for a winemaker. Using the opportunity as a ticket to get back home, he took the job and within a year had moved on to work with other brands, including Pillsbury Wine Company, launching his own label in 2013 under the name Rune Wines. Lured by the rolling grasslands of Sonoita, Callahan left Willcox to begin building his own tasting room, which opened in 2015. While he was originally able to make his wines at Pillsbury Wine Company, the move led him to transfer his production to Deep Sky Winery, for whom he was also consulting and making wine.

For the Rune wine label, Callahan sources a lot of his fruit from vineyards all over the state, which allows him to show how the different microclimates express themselves in the bottle. His focus is on varietal releases, though his forthcoming second label, Brigand, will feature unique blends.

Callahan takes a low-intervention approach with his wines, using native yeasts to ferment small lots of varietal wines with an emphasis on Rhône varieties. He describes his method as "high-risk winemaking" since he picks grapes that tend to have higher pH, cold soaks them for two days with dry ice, and uses wild fermentation following carbonic maceration for all red wines. But whatever his risk tolerance, the wines seem to be better for it. In general, Callahan pushes for more concentration and depth in his wines. The result is a distinctive selection of wines that combine Old-World structure and earthiness with New-World fruitiness and vibrance.

Try the clay-aged Picpoul Blanc, the richly complex Grenache, or the distinctive Syrah from Eric Glomski's Colibri Vineyards in the Chiricahua Mountains.

Sonoita Vineyards
Elgin
sonoitavineyards.com

Having opened its doors in 1983, Sonoita Vineyards, founded by Arizona wine pioneer Dr. Gordon Dutt, has the distinction of being the oldest continually operating vineyard in Sonoita. Following his exploratory success with planting experimental vines in Sonoita in 1979, Dutt opened the winery a few years later.

In the early years, Dutt compared the soils in Sonoita to Burgundy's Côte D'Or with its balance of red, weathered clay, and a deep layer of calcium carbonate. His initial assumptions were that the grapes for red wine would be overwhelmed by the intense Arizona sun, producing wines with poor color and low acidity. In fact, the opposite was true, as the red grapes instead developed thicker skins and sturdy phenolic structure.

Back then, Dutt produced about 300 gallons (1,135 liters) of wine. Today, Sonoita Vineyards is producing more than 10,000 gallons (37,854 liters) annually. The picturesque winery site overlooks the scenic hills of the Sonoita region, including 30 acres (12.1 hectares) of vineyard, primarily planted to Sauvignon Blanc, Cabernet Sauvignon, Merlot, Pinot Noir, Mission, Sangiovese, and Syrah. The winery continues to remain in the family, with Dutt's granddaughter, Lori Reynolds, serving as winemaker and vineyard manager.

In the interest of quality: the Arizona Vignerons Alliance

As a burgeoning wine region Arizona has chartered a rapid course over the past two decades, earning itself a seat at the international wine table for both its quality and consistency. Much of this can be attributed to a favorable climate and good grape-growing and winemaking practices. Still, it also helps that there's a great deal of organization and cohesion within the industry, which has helped propel it forwards. The statewide Arizona Wine Growers Association plays a crucial role in representing and promoting the industry, educating consumers about Arizona wine, and aiding legislative initiatives that support the industry. Regional organizations such as the Verde Valley Wine Consortium aid in promoting and marketing specific wine regions to consumers. In 2015, a small contingent of the state's top producers launched the Arizona Vignerons Alliance to establish a system for recognizing a standard of quality for wine.

At its core, the organization was created to support growers and winemakers within the state whose common goal is to establish Arizona as a reputable wine-producing region. The concept is much like the Vin de France classification or the Verband Deutscher Pradikatsweingüter (VDP), which certifies Grosses Gewächs status for German dry wines. Though the Vin de France is the least strict of French wine certifications, it still establishes a minimum standard of quality and requirement of origin. As with the German VDP, the certification process is entirely voluntary, allowing all Arizona producers, for a small fee, to submit their wines for evaluation by a panel of industry professionals without any direct connection to Arizona wine.

The primary stipulations of any wine selection are that it be 100 percent *Vitis vinifera*, 100 percent Arizona grown, and 100 percent Arizona produced. The threshold by which wines achieve certification is solely based on the requirement that they are without any perceptible sensory flaw and that panelists would gladly recommend them to general consumers. Wines approved by the AVA are labeled with a seal that guarantees both origin and quality for consumers, acting as a certificate to help instill confidence in the Arizona wine they may be considering. The idea is along similar lines to the DOCG wines of Italy, which bear a paper band containing a serial number.

The founding members include Kent and Lisa Callaghan of Callaghan Vineyards, Todd and Kelly Bostock of Dos Cabezas WineWorks, Maynard and Jennifer Keenan of Caduceus Cellars, and Rob and Sarah Hammelman of Sand-Reckoner Vineyards. As more Arizona producers have come to recognize the organization's potential for pushing the industry forward, nearly 30 wineries have entered their wines into the certification process.

With each passing vintage and evaluation, the organization has collected relevant data on the wines entered that can help point to evidence of which varieties are showing the most promise and in which specific regions. The hope is that this ongoing catalog can help guide growers and producers in their efforts both in vineyard and cellar.

Although altruistic in its motives, the organization has met with skepticism from some in the industry. There are concerns that the certification process creates a misconception of elite status. Still, while some discord is to be expected, the fact that the concept is well-established in many other regions across the world demonstrates that the organization is on to something. The fact that more than a third of Arizona producers have opted to participate in the evaluations since the AVA was launched suggests that it is gaining traction. What is true is that the alliance is the first of its kind established among the states of the Southwest, which, if nothing else, cements Arizona as a leader in setting the pace for quality among the respective regions as they move into the future.

WILLCOX AVA

Bodega Pierce/Saeculum Cellars

Willcox

bodegapierce.com

Though many may know Michael Pierce as the Director of Viticulture and Enology at the Southwest Wine Center at Yavapai Community

College, it's important to note that he's more than a school administrator. In fact, when it comes to Arizona wine, he's heavily involved. Since 2010, he has been a partner with his parents in Willcox-based Bodega Pierce and Rolling View Vineyard.

While studying electronic media and visual communication at Northern Arizona University, the younger Pierce had taken up brewing and winemaking as hobbies. After graduation, he decided to take online courses in viticulture and enology through the UC Davis extension. In 2007 he switched to Washington State University's program. At the same time, his father, Dan, a former advertising executive and graphic designer, had also taken a serious interest in viticulture. So he joined his son in his studies, simultaneously enrolling in the UC Davis and WSU viticultural programs and investing some time working at Owen Roe winery in Washington's Yakima Valley.

Following his enology coursework, Michael worked harvest jobs in New Zealand, Oregon, and Tasmania, before returning to Arizona and accepting a winemaking job in 2010 at Arizona Stronghold Vineyards. In 2011, his parents purchased an 80-acre (32.4-hectare) estate on the Willcox Bench, which had already been planted in 2005 by Jon Marcus of Echo Canyon Vineyard & Winery. For Dan, the vineyard has been a way to return to the farming that is the family's heritage. According to him, the Pierces have been farming in North America since before the states were united. Today, the Rolling View Vineyard has 36 acres (14.6 hectares) in production and sits at 4,300 feet (1,311 meters) with well-draining gravelly and sandy loam soils. Dan manages the 18 different varieties planted in the vineyard, including red and white Rhône and Bordeaux varieties, Petite Sirah, Tempranillo, Malvasia Bianca, Sangiovese, and Chardonnay. Both the elder and younger Pierce are particularly fond of Graciano and Pinot Noir, which they contend stands on its own merits as representative of the Willcox terroir.

When Four Eight Wineworks (see page 177) opened in 2013, Michael became one of its first tenants with his family brand, Bodega Pierce, along with his own private label, Saeculum Cellars. In 2014, he became Director of Viticulture and Enology at Yavapai Community College. While his full-time job demands the majority of this time, with his family managing the vineyards in Willcox, and his ability to focus on winemaking during harvest season, Pierce has found his place as an integral part of the Arizona wine story.

Try the beautifully balanced Malvasia Bianca, the brawny yet restrained Graciano, and, of course, one of the few Arizona Pinot Noirs on the market.

LDV Winery
Scottsdale
ldvwinery.com

Out beyond the eastern stretches of Willcox, in the volcanic soils of the Chiricahua Mountains, LDV Winery has staked a claim as an Arizona Wines outlier. Here, married couple Curt Dunham and Peggy Fiandaca transitioned their careers as consultants in urban planning to winemaking, a pursuit led by their many years of collecting wine. In 2008, the two bought a 40-acre (16.2-hectare) property and went to work planting 13 acres (5.3 hectares) of vineyard, primarily to Rhône varieties such as Grenache, Syrah, and Viognier, but also Petite Sirah.

Dunham likens the terrain to that of the western slope of the Sierra Nevada mountains in Amador County, California. The vineyards are planted at an elevation of 5,000 feet (1,524 meters), shielded from inclement weather by the soaring 10,000-foot (3,048-meter) peaks of the mountains behind it. The area draws its water supply from the plentiful Willcox Playa aquifer as well as from seasonal rainfall, which is rapidly drained away by the rocky, volcanic, and granite soils on which it is planted.

Today, LDV produces about 3,000 cases of primarily varietal wines reflective of its unique estate microclimate. While it welcomes visitors to the location in the southeastern quadrant of Arizona, it also opened a Scottsdale tasting room in 2018 to put the wines within convenient reach of the Phoenix-metro population. The wines are typically rich and robust, with broad structure offering an alternative to the more restrained styles of other southern Arizona producers.

Try the Viognier, which is surprisingly mineral-driven and redolent with lemongrass, or the Grenache, richly concentrated, with notes of tobacco and cherries on the nose and palate, and a long finish.

Pillsbury Wine Company
Willcox
pillsburywine.com

As an early advocate of Arizona wine, it's a toss-up as to whether Sam Pillsbury is more noted for his second career as an Arizona wine producer or his first career as a filmmaker. Born in Connecticut but brought up in New Zealand, Pillsbury discovered Arizona wine while on a film project in the early 1990s, where he fell in love with his future wife and made the Grand Canyon state one of his three homes (he currently divides his time between Willcox, Los Angeles, and New Zealand).

Having tasted a Callaghan Vineyards Chardonnay from Al Buhl's vineyard and been impressed with its Burgundian qualities, Pillsbury planted 20 acres (8.1 hectares) of Malvasia, Grenache, Mourvèdre, Syrah, and Petite Sirah in partnership with Buhl on adjacent property. In 2006, having sold this original plot, he purchased a 100-acre (40.5-hectare) property, right across the road from the Al Buhl Memorial Vineyard on the Willcox Bench. He initially planted the same varieties as on his initial plot, with the addition of Viognier, Chardonnay, Pinot Gris, Roussanne, Sangiovese, and Symphony, a white cross between Muscat of Alexandria and Grenache Gris. Though once a part of the Dos Cabezas WineWorks brand with Buhl, Pillsbury eventually struck out to create his own label. At first, Eric Glomski of Page Springs Cellars made Pillsbury's wines at his facility in the Verde Valley. But in 2014, Pillsbury built his own winery in Willcox with the help of upstart winemaker James Callahan of Rune Wines. Today, with a total of about 29 acres (11.7 hectares) in production, Pillsbury sells nearly half of his grapes to other wine producers while keeping the other half for his own brand. Though his production has moved to Willcox, Pillsbury continues to operate his Verde Valley tasting room in Cottonwood.

A big believer in a wine's ability to express authentically the place from which it comes, Pillsbury picks his grapes a little early to prevent over-ripeness, and uses wild yeasts for fermentation in neutral barrels. Pillsbury has never shied away from an opportunity to push the Arizona wine message forward. Were it not for his confidence in what he and others are doing, he might not have been given a second glance by FnB's Pavle Milic when he strolled up to the newly opened restaurant in 2009 with a few of his wines. This pivotal encounter would lead Milic to launch an Arizona-only wine list at the James Beard Award-winning restaurant. To date, Pillsbury is one of the most awarded producers in the state, and with good reason. His wines testify that hard work in a desolate place can produce finesse and balance.

Try the mineral-driven Viognier, the earthy Grenache, or the food-friendly Roan Red blend.

Sand-Reckoner

Tucson
sand-reckoner.com

Third-century mathematician and scientist Archimedes coined the term sand-reckoner. It was a theoretical calculator to determine the size of the universe through the counting of the number of grains of sand that could fill it. This existential concept reflects how Rob and Sarah Hammelman see the sandy loam soils of Willcox combined with the myriad factors that must come together to produce a wine that authentically represents its place. More simply, Sand-Reckoner wines are delicious and taste like Arizona.

Rob Hammelman first discovered his love of wine while crushing a harvest at Callaghan Vineyards in 2000. During this foundational experience with Callaghan, Hammelman was struck with the realization that the wines were intrinsically representative of their place.

In search of more in-depth experience, he headed to the Southern hemisphere to pursue a graduate diploma in Oenology at the University of Adelaide in South Australia. There, he deepened his understanding of Rhône varieties in the New World while working for celebrated producer Hewitson, in one of the world's oldest vineyards still growing, the Old Garden Mourvèdre, planted in 1853. After gaining his diploma, he moved to Colorado to make wine for Two Rivers Winery in Grand Junction, with varieties such as Riesling grown at elevations of 6,000 feet (1,829 meters). Here he met Sarah, a Colorado native who was working a harvest at the winery and later worked a harvest at California's Etude winery. Before returning to Arizona, Rob secured an opportunity to work with Rhône varieties in the grapes' native home with an apprenticeship at Chateau St. Cosme in Gigondas, France, for notable winemaker Louis Barroul. With Sarah alongside him, the experience of working in a 2000-year-old cellar with Grenache vineyards dating back more than 100 years brought Hammelman's vision for his own winery into focus.

In 2010, the Hammelmans moved to Willcox to plant a vineyard and launch Sand-Reckoner. They purchased a 12-acre (4.9-hectare) farm initially planted with a small vineyard in 1997 and located between the Chiricahua and Dragoon Mountains outside of Willcox. The soils of the Sand-Reckoner Vineyard, which sits at 4,300 feet (1,311 meters), are well-drained sandy loam layered atop a limestone subsoil not four feet (1.2 meters) beneath the surface. Here, the Hammelmans grow Malvasia Bianca, Nebbiolo, Sangiovese, Sagrantino, Syrah, Petit Verdot, and Zinfandel. They also source fruit from other vineyards throughout the region including the Chiricahua Ranch Vineyard, located in Willcox at 4,530 feet (1,381 meters) on granite-sandy loam; Bloomoon Vineyard, which is situated on the granitic and volcanic fan terrace of the Chiricahua Mountains at 4,697 feet (1,432 meters); the Red Tree Ranch Vineyard on a remote site at 4,976 feet (1,517 meters) in the foothills of the Chiricahua Mountains with granite- and gneiss-derived sandy loam; the Rhumb Line Vineyard at the northern edge of the Willcox Bench at 4,300 feet (1,311 meters) with loam and deep alluvial soils rich in clay and limestone; and the Cimarron Vineyard of Dos Cabezas WineWorks at 4,300 feet (1,311 meters) in Willcox.

Sand-Reckoner wines are bold and generous, as you might expect from the New World, but they also have earthy finesse and layered depths resonant of Old-World styles. Some might say they reflect the Hammelmans' previous experiences in winemaking regions across the world. But it's perhaps more accurate to say that they are simply a reflection of Arizona itself. Try the Vermentino from the Chiricahua Ranch Vineyard, the crisp but berry-forward Grenache rosé, and the earth-driven Sangiovese.

VERDE VALLEY

Arizona Stronghold Vineyards
Cottonwood
azstronghold.com

Arguably the winery that put Arizona wine on the map, Arizona Stronghold Vineyards emerged from a collaboration between Maynard James Keenan and Eric Glomski in 2007. They had purchased the Dos Cabezas Vineyard, originally planted by Robert Webb in 1983, and taken over by Al Buhl in 1995, to start up the Dos Cabezas brand. They founded Arizona Stronghold, which derives its name from the location of vineyards that sit between the Dragoon and Chiricahua mountains; the most prominent edifice of the Dragoon range is called the Cochise Stronghold. Historically, this rugged mountain site served as the Apache settlement in which the storied Chiricahua Apache Chief, Cochise, made his camp in preparation for the Battle of Apache Pass in 1862 (see box, p. 132).

From the beginning, Arizona Stronghold sought to create quality wine at a value-driven price point with retail market appeal and distribution

170 THE WINES OF SOUTHWEST U.S.A.

outside state borders. Glomski and Keenan hired Tim White from Virginia as winemaker, and together the three succeeded in steadily growing the brand while simultaneously making a name for Arizona wine. In 2014, the partnership split, with Keenan retaining ownership of the Dos Cabezas Vineyard—which he renamed Al Buhl Memorial Vineyard—while Glomski kept the winery, and a second vineyard, Bonita Springs. As the brand continued to grow, Glomski enlisted family help from his father, Terry Glomski, and stepfather, Rod Young, to run Arizona Stronghold while he focused on Page Springs Cellars. Today, Glomski has relinquished full ownership to his family members.

More than half of the grapes used in Stronghold wines are sourced from the company's Bonita Springs Vineyard, with most of the remaining grapes sourced from other vineyards in the Willcox area. They are brought up to the Verde Valley winery, which is housed in an old furniture factory in Camp Verde. Arizona Stronghold wines are generally easy-drinking and accessible, reflective of the Arizona growing conditions, offering bright fruit character and earthy undertones. Their second label, Provisioner Wine, serves as a broad-appeal, value-driven retail selection of wines. Try the Arizona Stronghold Mangus Red, a rustic Super-Tuscan-style blend of Cabernet Sauvignon, Merlot, and Sangiovese, or the minerally Dala Chardonnay.

Burning Tree Cellars
Cottonwood
burningtreecellars.com

Located in quaint Old Town Cottonwood, Burning Tree is a partnership between notable Page Springs Cellars winemaker Corey Turnbull and Mitch Levy, a graduate of the Southwest Wine Center. This boutique winery offers limited-release offerings from vintage to vintage, based on grapes sourced from select Arizona vineyards. Rhône varieties such as Syrah and Mourvèdre are consistent, with nuance and regional character from Eric Glomski's Colibri Vineyards, as is the Viognier, which is sourced from the Deep Sky Vineyards in Willcox.

Caduceus Cellars and Merkin Vineyards
Jerome
caduceus.org / merkinvineyards.com

On paper, Maynard James Keenan is the owner and co-winemaker of Caduceus Cellars and Merkin Vineyards, owner and founder of the

Arizona-based winemakers co-op Four Eight Wineworks, and a founding board member of the Arizona Vignerons Alliance, a quality-control and winegrowers' data collection collective. But those are just a few of his jobs. In non-wine circles, Keenan is a rockstar. Literally. For the better part of three decades, he has been the Grammy Award-winning vocalist for the internationally revered music acts Tool, A Perfect Circle, and Puscifer. But while his life as an internationally known musician still has a few chords left in it—Tool released a long-awaited album in 2019, the first in more than a decade—in many ways, Arizona wine is the future for Keenan.

Originally from Ohio, Keenan grew up in Michigan and spent his early adulthood in the U.S. Army before moving to Los Angeles to pursue a music career. His alt-rock band Tool, formed in 1990, swiftly rose in fame and international acclaim. Keenan's music career catapulted him into a lifestyle that would not only allow him to see the world, but also appreciate its many cultures, foods, and wines. After living the rock and roll life in L.A., he discovered Arizona's Verde Valley, making it his home in 1995. Inspired by the many wine regions he had visited around the world, Keenan was struck by the similarities the Verde Valley bore to areas of Spain, Italy, and France. In 2002, he planted his first vineyard, an endeavor that would quickly succumb to winter kill. Recognizing the need for frost protection and careful site selection, he planted again, this time with encouraging success.

Today, Keenan owns and manages seven vineyards in both the Verde Valley and the Willcox AVA. The two-acre (0.8-hectare) Judith's Block sits at 4,900 feet (1,494 meters) on Cleopatra's Hill in Jerome and includes about a half-acre each of Nebbiolo, Aglianico, Tempranillo, and Malvasia Bianca. This block is unique for its steep slope and hand-cut terraced rows all upon volcanic topsoils that cover a limestone-caliche layer beneath. The 2.5-acre (1-hectare) Marzo Block is located along Oak Creek in Cornville at an elevation of about 3,500 feet (1,067 meters). Here, Sangiovese, Cabernet Sauvignon, and Syrah are planted on topsoils of sandy clay-loam river sediment layered underneath with volcanic rock that is peppered with calcareous sediment. The 6-acre (2.4-hectare) Agostino Block is planted with Mourvèdre, Albariño, Lulienga/Lignan Blanc, and Malvasia Bianca and is considered the most challenging vineyard when it comes to spring frosts (the addition of an orchard fan has helped). Originally planted in 2012, the 30-acre (12.1-hectare) Eliphante Block near Cornville includes nearly two dozen varieties, with an emphasis on Italian cultivars such as Nebbiolo, Barbera, and Aglianico, but also includes French and Spanish varieties such as Cabernet Sauvignon, Syrah, and Tempranillo. The Southwest Wine Center Block is part of the viticulture and enology program at Yavapai Community College. It has grown from one acre (0.4 hectares) of Negroamaro, which Keenan donated, to 13 acres (5.3 hectares) of different varieties planted by students of the program. Keenan also leases a small parcel of Barbera near the original Marzo Block. Finally, the historic Al Buhl Memorial Vineyard was initially planted in 1983, and 20 of the original acres (8.1 hectares) are still in production today. The entire 80-acre (32.4 hectares) site has 60 acres (24.3 hectares) planted with nearly two dozen varieties, including Barbera, Grenache, Malvasia Bianca, and Sangiovese.

Keenan is not naive to the fact that being a rockstar doesn't exactly give you unquestioned credibility in the wine world. He is not the first celebrity to back a wine label, but Keenan isn't the sort of person to buy a wine label simply to add it to a list of hobbies. Nor is he the type to casually enter into a new endeavor. Instead, he hand-picked a team, which included his co-winemaker Tim White, who had previously made the wines for Arizona Stronghold, and vineyard managers Chris Turner, for the Verde Valley vineyards, and Jesse Noble, for the Willcox vineyards. Though White signed on to help oversee wine production, it is not a solo effort as Keenan has adjusted his extensive touring schedule to be hands-on during the harvest season. In addition, Keenan has spent the past two decades seeking consultation from some of the world's most renowned enologists and viticulturists, including Peter Gago of Penfold's in Australia and Taras Ochota of Ochota Barrels, as well as Luca Corrado of Piedmont's Vietti, and celebrated pruning consultant Alessandro Zanutta.

It would be one thing if Keenan were just in it for himself, but from the very beginning the momentum with which he has energized the Arizona wine industry has been both purposeful and collaborative. He has played a pivotal role in the creation and growth of the Southwest Wine Center, investing in the school's first vineyard planting as well as in scholarship opportunities for would-be enology and viticulture students. Through the Arizona Vignerons Alliance, he has helped to spearhead the push for quality wine throughout the state. Through his wine incubator, Four Eight Wineworks, he has assisted new producers in need of resources to launch their brand. And he has tactfully leveraged his international fame to act as a mouthpiece for getting the word out about the burgeoning wine region. To those within the industry, Keenan's presence during harvest season is palpable and obliging, showing up with extra harvest bins, offering up equipment, or helping to secure more fruit for an upstart producer in need. Indeed, his self-imposed mission from the beginning has been that "Arizona would not just be seen as an emerging wine region, but sought after."

It helps in Keenan's case that the wines speak for themselves. The offerings of both Caduceus Cellars and Merkin Vineyards are complex and thought-provoking, offering elements that are varietally correct, but also expressive of place. Caduceus Cellars functions as Keenan's more premium brand, featuring small-batch blends and varietal releases. Try the Kitsuné, a sunny Sangiovese from Cochise County (Willcox AVA) with bright red fruit and earthy, umami undertones, or the brawny yet elegantly structured Nagual del Judith Aglianico. Merkin Vineyards selections are equally appealing but more approachable in price. Try the classic Provence-style Chupacabra Rosa, a cheerful rosé blend of Grenache, Syrah, and Mourvèdre.

Educating Arizona: the Southwest Wine Center

Clarkdale; southwestwinecenter.com

Just south of the town of Clarkdale on the way west to Jerome, in the heart of the Verde Valley, you can stand on an expansive patio and gaze over more than a dozen acres (4.9 hectares) of vineyards and into the southern bend of the Black Hills mountains. What you might not realize is that this used to be a racquetball gym. Not the vineyards, the building. Before 2010, every square inch of this state-of-the-art building with built-in sustainability features, solar lighting, and an advanced rainwater catchment system, used to be a recreational facility. Today, the site is home to the Southwest Wine Center of Yavapai Community College, one of only a handful of programs in the country offering an all-encompassing education on viticulture, winemaking, and the wine business.

At its inception in 2010, the idea for the program was to teach students real-life lessons about growing grapes—from vine to glass. Yavapai's program is modeled after the University of California, Davis program, which teaches students the business of producing and marketing wine in addition to viticulture and enology. Students entering the program can opt for a one-year certificate in either viticulture or enology, or a two-year degree in both.

A donation from Caduceus Cellars owner Maynard James Keenan allowed students to plant their first acre (0.4 hectares) of a vineyard at the site near Clarkdale (they planted Negroamaro). As the program gathered momentum over the following years, the need to include a wine-production facility became a reality, and in 2013 the racquetball courts were transformed into a full-scale winery. What used to be the racquetball bays became a tank room, a barrel room, and a sleek, inviting tasting room. By 2014, the first class was graduating from the program with degrees in applied science in viticulture and enology.

Vineyards at the Southwest Wine Center in the Verde Valley

In the past decade, the original one-acre (0.4 hectare) planting has grown in size to become 13 acres (5.3 hectares), all of which is planted, managed, and harvested by students who then produce their own wines, design their own labels, and formulate their own business plans for their own potential winery. The overall case production from the Southwest Wine Center is 2,000.

In 2015, the program opened its own tasting room, allowing the general public to taste the wines made by students. The tasting room is staffed by students who learn direct-to-consumer marketing as part of the college's curriculum and have a chance to hone their customer service skills.

According to Michael Pierce, the Director of Viticulture and Enology for the Center, students get a clear, unbiased look at how a wine business is run, learning what it takes to do everything from planting a vine to understanding the licensing process and managing a wine club.

The course offerings at the Southwest Wine Center have attracted students with a mix of ages and life experiences—the average age of students in the program is 48. There are professionals looking for a career change studying alongside twenty-somethings looking to get in on the ground floor of a new career. By 2019, enrollment was at capacity, with 105 students entering in the fall.

Many graduates of the program have been hired by other Arizona wine producers including Chris Brinkmeyer at Pillsbury Wine Company, and Tom Messier with Rune Wines and Deep Sky Winery. Other graduates have launched their own wine brands in Arizona, including Valerie and Daniel Wood of Heartwood Cellars, and Mitch Levy of Burning Tree Cellars. Though perhaps not the original intent of the program, its steady stream of graduates has undisputedly become a boon for the Arizona wine industry, helping to fulfill its need for a skilled, educated workforce.

In addition, the Southwest Wine Center is equipped with a lab for wine analysis. This facility will eventually serve as a resource for Arizona's wineries to use rather than sending samples to California, and become another source of revenue for the school. The college has also partnered with the University of Arizona's College of Agriculture and Life Sciences to create a data repository, which will help catalog vineyard data, grape variety successes, and best practices. This will eventually become a community resource for state-specific information about grape growing and winemaking. In 2020, it launched an additional Certificate in Brewing Technology to support many of the craft beverage industries in the region.

Today, the Southwest Wine Center is not only an educational hub for aspiring viticulturists and enologists and a destination for wine enthusiasts out to sample the fruits of student labor, but also serves as a model for other would-be wine university programs in emerging wine regions around the country.

Chateau Tumbleweed

Clarkdale

www.chateautumbleweed.com

Chateau Tumbleweed is perhaps the best example of a winery bred out of Arizona's modern wine industry. Each of its four owners, all originally from the west coast, has a unique experience with some of the state's top producers. Joe Bechard, a journalist-turned-winemaker, earned his chops at Page Springs Cellars before making wine at Alcantara Vineyards and Winery and helped grow the Merkin Vineyards label for Maynard James Keenan. His wife, Kris Pothier, label artist and marketing guru, worked tasting room, shipping, and sales for Page Springs Cellars and Caduceus Cellars. Jeff Hendricks, vineyard manager and graphic designer, also managed the vineyards for Page Springs Cellars. His wife, Kim Koistinen, is an accountant who managed operations for Page Springs Cellars as well.

The brand launched its first vintage in 2011 with the help of nearby winery Alcantara Vineyards and Winery, before becoming the first participant in the Four Eight Wineworks incubator program. Benefitting from the Four Eight Wineworks shared space and equipment, the team invested time in releasing some of the state's most nuanced and expressive wines while also honing their marketing and business strategy in anticipation of moving into their own space.

In 2014, Earl and Melinda Petznick of the long-established D.A. Ranch were in search of help in growing their own wine label. Taking note of the power team's steadfast work, they formed a partnership that helped finance the build-out and winery equipment for Chateau Tumbleweed in Clarkdale and included vineyard management by Jeff Hendricks for the D.A. Ranch Vineyard. By 2015, Chateau Tumbleweed had moved into its new premises. Currently, the winery produces about 5,500 cases, the majority of which is for Chateau Tumbleweed, with about 1,000 cases under the D.A. Ranch label.

The winery sources about 80 percent of its fruit from vineyards in and outside of the Willcox area, including Cimarron Vineyard, Carlson Creek Vineyards, Pillsbury Vineyards, the Al Buhl Memorial Vineyard, Sierra Bonita Vineyards, Deep Sky Vineyards, Rhumb Line Vineyard, Edge of the Vine Vineyards, Rolling View Vineyard, Colibri Vineyards, House Mountain Vineyard, and Dos Padres Vineyards.

The wines are clean and focused, with true varietal character for their varietal wines and elegant, finessed structure for their blends. The winery's favorite varieties to work with include Aglianico, Vermentino, Picpoul Blanc, Graciano, Sangiovese, and Montepulciano.

Chateau Tumbleweed exudes every bit of the meaning behind its name. Pothier describes the moniker as representing its four Gen-X owners, who found their way to Arizona and got stuck like a tumbleweed in a barbed-wire fence. They are serious about making excellent Arizona wine but do not take themselves too seriously. Their labels are silly, fanciful, and often with a tongue-in-cheek meaning, as is the use of the word "chateau" in the brand name. You won't find any formal French châteaux in Arizona, but you will discover châteaux-worthy wine in everything this team produces.

Though the use of the lowly, ambling tumbleweed in their name belies a certain casual, laidback state of mind, Chateau Tumbleweed has lit a fire under this cliché to become more of a rocket blazing a trail into Arizona's wine future. The team have checked all the boxes in terms of taking advantage of their network of experiences, using an incubator program to its fullest value before carving their own place in the market, while simultaneously raising the bar for overall quality in Arizona. Perhaps that sounds like a grand statement, but when you taste these wines, you begin to get a sense of its verity. Try the beautifully complex Edge of the Vine Sangiovese, with tart red cherry and a pleasingly dusty mouthfeel, or the rich, dark blue fruit notes of the House Mountain Syrah, offering a bold palate that finishes with restraint and finesse.

Four Eight Wineworks
Clarkdale
four8wineworks.com

Realizing yet another way to fulfill a need in the industry, in 2013 Maynard James Keenan launched Four Eight Wineworks as a way to give young upstart producers a chance at success. The operation functions as an alternating proprietorship, in which the participating wineries each have their own federal permit. Tenant wineries alternate the use of the facility's production equipment, each occupying its own designated footprint, supplying its own fruit, managing its own production and packing materials, and taking responsibility for all necessary reporting, labeling, and taxes.

The name refers to Arizona's position as the forty-eighth state to enter the Union. Housed in an old bank building in downtown Clarkdale, the facility is the first co-operative in the northern part of the state, serving as an incubator for up-and-coming producers to explore winemaking, while also developing a business plan for their brand—all without the added expense of building a bricks and mortar winery. The result is an ideal environment for creativity and experimentation for quality winemaking. At Four Eight, winemakers have shared use of vital equipment, from a press and de-stemmer to steel tanks and a bottling line. The space includes a tasting room, where each participating producer has its own bonded space for selling its wines. Would-be participants must submit an application and, if accepted, pay a fee to use the space based on the amount of wine produced. It's a concept widely employed in wine regions all over the world, but rarely seen, as of yet, in America's smaller, burgeoning regions. To date, the project has helped launch

Bodega Pierce and Saeculum Cellars, The Oddity Wine Collective, Heartwood Cellars, and Chateau Tumbleweed.

Javelina Leap Vineyard & Winery
Cornville
javelinaleapwinery.com

In the history of Arizona wine, Javelina Leap is an important player. It was started by husband-and-wife team Rod and Cynthia Snapp in Cornville in 1999, when winemaking in the Verde Valley was virtually non-existent, but they were convinced they could make it work. Named for the small, native wild boar-like creature of the American southwest, Javelina Leap manages a 10-acre (4-hectare) estate vineyard overlooking the verdant Oak Creek greenbelt at about 3,800 feet (1,158 meters). Planted in primarily volcanic soils, the vineyard includes Cabernet Sauvignon, Merlot, Tempranillo, Sangiovese, and Barbera. The Snapps also buy fruit from a few other Arizona growers. Try the Legacy Zinfandel, one of the Snapps' celebrated varieties, and the rustic, tobacco-scented Tempranillo.

The Oddity Wine Collective
Jerome
www.theodddtywinecollective.com

A partnership of three graduates from the Yavapai Community College Southwest Wine Center, the Oddity Wine Collective was the first concept to ease into a spot at the Four Eight Wineworks co-op program when Chateau Tumbleweed was ready to move out. Finding each other through coursework during their time at Yavapai, business partners David Baird, Briana Nation, and Aaron Weiss recognized individual strengths that, together, could create a viable wine brand. By 2014, they had formalized a business plan and become officially incorporated. In 2015, they delivered their first vintage. Though all three partners in the collective still maintain their day jobs in the Arizona wine industry, the hope is to eventually establish their own space. Currently, the team sources fruit from a handful of different vineyards throughout Arizona's wine regions, including Willcox fruit from Cimarron Vineyard and the Al Buhl Memorial Vineyard as well as Sonoita's Deep Sky Vineyards. Try the Minotaur Mourvèdre or the Unsanctioned, a 50/50 blend of Sangiovese and Petite Sirah.

Page Springs Cellars

Cornville
pagespringscellars.com

Page Springs Cellars is one of the foundational producers of the Verde Valley. Founder and owner Eric Glomski operates it as a family venture, making an honest effort to create authentic wines that represent the Arizona landscape—and by all accounts, his hard work has paid off. Though originally from the Chicago suburbs, Glomski has spent the better part of his life claiming Arizona as his home. He attended Prescott College to study riparian ecology during which time he took up the pastime of making wines from local apples. But what began as a hobby soon commanded his full attention.

In 1996, he headed out to California's Santa Cruz mountains to take a cellar job at David Bruce Winery. Over the next six years, he would work his way up to co-winemaker. By 2002, he was ready to move back to Arizona, where he took a job as winemaker for Echo Canyon Winery. In 2004, with the help of his family, Glomski launched Page Springs Cellars. Owing to his collaborative nature, he became involved with several other upstart labels, helping to make their wines, or letting others use his facilities. In 2007, Glomski partnered with Maynard James Keenan to create Arizona Stronghold. But in time, these other projects began to overshadow the work he really wanted to do, which was to focus on the terroir-driven wines for Page Springs Cellars. By 2015, he and Keenan had dissolved the Arizona Stronghold partnership leaving Glomski with the wine brand and their estate, Bonita Springs Vineyard, while Keenan assumed ownership of the original estate Arizona Stronghold vineyards in Willcox.

Glomski has since handed ownership of Arizona Stronghold over to his father and stepfather, Terry Glomski and Rod Young, who continue to grow the brand today. Page Springs Cellars has become Glomski's central focus. For Glomski, his best days are those spent in one of the four vineyards he owns and manages. He has given day-to-day winemaking responsibilities to Corey Turnbull, who also owns Burning Tree Cellars (see p. 170) with Mitch Levy.

The first estate vineyard, Home Vineyard, was planted in 2004 and currently includes 4.5 acres (1.82 hectares) of Marselan, Montepulciano, Grenache, Traminette, and Seyval Blanc. The vineyard sits at 3,450 feet (1,052 meters) on sandy clay-loam soils with a calcareous layer underneath. House Mountain Vineyard sits a little higher at 3,550 feet

180 THE WINES OF SOUTHWEST U.S.A.

(1,082 meters) on the warmer, western slope of House Mountain volcano. The 8 acres (3.2 hectares), which were planted in conjunction with the Dancing Apache Ranch, include Petite Sirah, Pinot Noir, Syrah, Teraldego, Grenache, and Counoise. The third Verde Valley planting is the 9-acre (3.6-hectare) Dos Padres Vineyard, which lies across Oak Creek at elevations between 3,400 and 3,650 feet (1,036–1,113 meters). The chalky clays include chunks of volcanic stones as well as ash and limestone deposits. Here, Glomski grows Syrah, Carignan, Aglianico, Barbera, Grenache Blanc, Roussanne, Vermentino, and Malvasia Bianco.

The fourth and final vineyard is located further south, east of the Willcox AVA in the Chiricahua Mountains. The unique mountain vineyard, originally planted in 2000, sits at about 5,200 feet (1,585 meters) and has rhyolitic, volcanic soils. Here, Glomski grows a few different clones of Syrah and Grenache as well as Mourvèdre, Counoise, Cinsault, Durif, and Roussanne. Glomski refers to the Colibri Vineyard as the Grand Cru of Arizona, if only to illustrate the quality of the fruit from this high elevation site. The vibrant fruit characteristics are framed by textural minerality and marked spice.

With his ecology background, the topic of sustainability is an important part of Glomski's farming philosophy. He maintains the vineyards with native grass cover crops, composts and reuses plant debris, and employs solar panels to supply electricity. In addition, the winery upholds a policy for employee wellness and fair wages.

The wines of Page Springs Cellars are characteristically Old World in style, boasting tertiary notes of earth and minerality, backed by elegant fruit and finessed mouthfeel. Try the Clone 474 Syrah from Colibri Vineyard next to Clone 471 Syrah (whole cluster) from House Mountain vineyard. The Counoise Rosé from House Mountain Vineyard offers delicate notes of summer berries and vervy lemon zest.

PHOENIX AREA

Garage-East
Gilbert
www.garage-east.com

A vibrant urban winery located in the suburban town of Gilbert near Phoenix, Garage-East is a serious wine producer that doesn't take itself too seriously. Case in point, the crowd favorite, and utterly gluggable

Breakfast Wine, a white wine blended with fresh fruit juice—whatever is local and in season—and carbonated. It's like a mimosa (champagne and citrus juice cocktail), but better, and well worth trying. The concept is the brainchild of Dos Cabezas owners Todd and Kelly Bostock, and former Gilbert Fire Department battalion chief Brian Ruffentine and his wife, Megan.

The winery sits at the far east end of Barnone, a self-proclaimed commercial "craftsman community", which is home to a coffee shop, a florist, a hand-made stationery store, a few chef-driven restaurants, a brewery, and this Arizona winery. The name is a play on the trendy term *garagiste*, though the winery's industrial structure does look a lot like a sleek mechanic's garage.

Wines are made on site from grapes harvested from vineyards in both Willcox and Sonoita. Bostock takes a light-hearted approach with the Garage-East wines with a mind towards making vibrant, fresh, and quaffable wines that are meant to be lively, young, and most of all, fun.

Almost all of the wines on offer are kegged, and served on tap. Guests order by the glass, by the bottle (or carafe), or do a flight of six wines served in a clever little wooden stand, which holds six 1-ounce (30-milliliter) glass "test tubes" of wine. All wines on offer can be bottled or canned from the tap. Try the aromatic Pinot Gris or the rich and provocatively complex Higley red field blend from the Bostocks' Cimarron Vineyard in Willcox. The Breakfast Wine is worth tasting for fun, as is the Sonoran Spritz, a concoction of house-made wine, bitter orange liqueur, and a dose of carbonation.

12

WHERE TO EAT AND STAY IN ARIZONA

SCOTTSDALE

Though Scottsdale could easily be characterized as a glitzy suburb of Phoenix, its rich agricultural heritage combined with its high desert Palm Springs feel makes it a single destination with a character all its own. Visitors from both east and west coasts, and everywhere in between, come to stroll the shops and galleries of Old Town, relax at its glamorous golf and spa resorts, and savor an ever-growing collection of dynamic dining options.

Stay

Bespoke Inn
Scottsdale
bespokeinn.com
This boutique hostelry in the heart of Old Town is walking distance from most shops, restaurants, and tasting rooms. Guests are given complimentary use of its bicycles, making it easy to stash the day's spoils in your bike basket and cycle back for a contemporary Italian dinner by chef Gio Osso at on-site Virtù Honest Craft.

Hotel Valley Ho
Scottsdale
hotelvalleyho.com
Set in the Arts District, this trendy vintage hotel dating from the 1950s is within walking distance of the multiple shops, restaurants, and bars

on Main Street. The entire property features a mid-century-modern design with rooms showcasing upscale furnishings. Enjoy a cocktail at the pool or the eclectic on-site restaurant.

Dine

The Farm at South Mountain

Phoenix
thefarmatsouthmountain.com

This sprawling farmstead is a collective of gardens, a park, event space, small businesses, and greenhouse all in one. The property offers a refreshing space for visitors to stroll around the grounds or stop into The Farm Kitchen, a cheerful counter-service lunch spot offering fresh sandwiches, salads, and soups using the seasonal bounty from the garden as well as other ingredients.

FnB Restaurant

Scottsdale
fnbrestaurant.com

Known for illuminating the agricultural bounty of Arizona, FnB chef and co-owner Charleen Badman showcases a veggie-dominant seasonal menu (there is a handful of meat and fish selections as well). Her prowess with regional produce has earned her the moniker "the veggie whisperer", as well as a James Beard Award for Best Chef Southwest in 2019. Her co-owner, Pavle Milic, acts as front-of-house general manager and seasoned sommelier, championing Arizona wine in tandem with international offerings. The restaurant's new adjacent FnBar also showcases those local vintages among global selections.

Pizzeria Bianco

Phoenix
www.pizzeriabianco.com

For something more casual, this cozy, brick-walled pizza kitchen serves as the little brother to acclaimed chef Chris Bianco's Tratto, offering authentic Neapolitan-style pizzas from a wood-fired Italian brick oven, along with classic Italian small plates such as Prosciutto-wrapped Fontina, or wood-roasted vegetables.

Southern Rail

Phoenix
southernrailaz.com

This cozy, local favorite marries Southern-inspired dishes with local, farm-fresh ingredients of the Southwest for an altogether soulful experience.

The Arizona standard: how FnB Restaurant helped put Arizona wine on the map

In recent years, Arizona's Phoenix, Sedona, Tucson, and Scottsdale have evolved to become individual hubs for a diversity of cuisines. Scottsdale, in particular, has become a hotbed for such remarkable dining destinations as Tratto, Talavera, Mowry & Cotton, and most notably, FnB. In 2009, chef Charleen Badman and business partner, beverage director Pavle Milic, opened FnB—which stands for "food and beverage"—quickly turning heads with an innovative farm-to-table menu showcasing seasonal bounty from local farms and other purveyors.

Both Badman and Milic had built admirable careers in their respective fields. As a self-taught chef, Badman had cut her teeth in Arizona's top restaurants, including Rancho Pinot and Café Terra Cotta, as well as at New York's famed Lobster Club, before becoming co-owner of the Greenwich Village restaurant Inside. Milic, a generously gracious host by nature, was raised in Queens by way of Colombia, his mother's native country. By 16, Milic, who had earned the strong work ethic of his Yugoslavian father, had fallen in love with the hospitality industry. In the years that followed, he would take on career-defining roles at Phoenix's Franco's Trattoria, Anne Rosenzweig's Arcadia in New York, Restaurant Budo, and Restaurant and Bar Angele, where he met one of his mentors, Bettina Rouas, in Napa Valley. During this time he had worked his way up from busboy to wine director. When Milic and Badman reconnected in Phoenix in 2007, the idea for FnB took root.

Since then, Badman's creativity and soulful approach to Arizona ingredients have earned her the coveted James Beard Award for Best Chef of the Southwest, in 2019. In tandem, Milic has earned a couple of James Beard nominations for Best Wine Program, a distinction particularly notable for the fact that in its formative years he curated the first Arizona-only wine program in the state (although as a back-up, they also had a "Plan B" wine list available upon request).

At first, the decision to commit so much real estate to Arizona wine wasn't an obvious one. Having built the wine lists at top restaurants and hotels in New York, Napa, and Arizona, Milic had access to share the whole world of wine to FnB guests, and when the restaurant first opened, he fully intended to do so. But when a casually-dressed middle-aged man strolled into the restaurant one afternoon before service and asked Milic to taste his wines, he realized Arizona wine was an idea he needed to consider. The man was Sam Pillsbury, filmmaker-turned-winemaker whose Pillsbury Wine Company had become one of the standard-bearers for Arizona wine. "Everyone thinks of Arizona as cowboys, cactus, and the Grand Canyon, not wine," says Milic. "But tasting these wines with Sam ignited my interest."

Considering FnB opened on barely a shoestring budget, the idea of designing a wine list with Arizona wine rather than highly-allocated, high-dollar wines from around the world also piqued Milic's interest. What it came down to, for Milic, was the fact that at its core, everything about FnB's menu was authentic to Arizona. "It just made complete sense. My business partner showcases everything local," says Milic. "It only made sense to bridge the gap between food and wine if guests could taste a little piece of Arizona sunshine in the bottle."

Milic began reaching out to different Arizona producers, requesting samples of wine to taste and familiarizing himself with the Arizona wine landscape. To his surprise, many of the wines he tasted were precisely what he had been looking for. "These wines were tasty and distinctive. They had earthiness and juiciness, redolent of the Old World," he says. Despite his gut impulse to serve only Arizona wine, Milic didn't entirely trust himself, nor the thirst of his customers. He began testing the waters with customers, bringing them a glass on the house during the meal without telling them where it was from to gauge their feedback. "There's nothing more democratizing than drinking something without baggage," says Milic. "There were no strings attached, and I soon found that people really liked the wines."

Within three weeks of opening in December of 2009, Milic authored the Arizona-only wine list. At first, people did question the move, but as anyone who has crossed paths with the utterly charming Pavle Milic knows, he could sell snow to a polar bear. But nothing could have prepared him or Badman for Valentine's Day of 2010. That morning, Milic opened the Sunday business section of the *New York Times*, to see his face along with a feature article about FnB. For him, there could not have been a better affirmation for everything that FnB was doing. "It was so interesting to experience the power of that publication," says Milic, who immediately began fielding congratulatory phone calls from friends all over the world. "It was a blessing."

Indeed, not three months after first opening its doors, FnB had received a resounding pat on the back for its brave, thoroughly local concept, right down to the Arizona wines it was proudly featuring. Soon after, *Food & Wine* magazine followed up with a piece naming FnB in its 10 Best Restaurant Dishes of 2010, mentioning the Dos Cabezas WineWorks Toscano as a perfect pairing for Badman's braised leeks with mozzarella and a fried egg.

In the following years, both Badman and Milic received countless accolades for their efforts, including multiple James Beard Award nominations, before Badman won hers in 2019. But despite the national attention, it is the support of the local Arizonans that has kept FnB alive. As the restaurant and its clientele have grown, so has the wine list, now with a considerable number of international wines among the continued strong presence of Arizona wines. For Milic, that evolution was a necessary next step: "Arizona wines deserve to be listed among the other iconic wines of the world. It's the only way they will move forward," he says. Milic knows that if he does his job right, Arizona wines will continue to show well: "We need to be proud of who we are and of what we are doing, and when there's that confidence, it's contagious."

In addition to his work at FnB, Milic has championed Arizona wine through his hosting of multiple blind tasting events for consumers and trade, and participation in the annual Arizona Vignerons Alliance Symposium. He writes a monthly column, "Pav on Juice", for *Phoenix Magazine*. In the past few years, his commitment to local wine has found him elbow-deep in planting a vineyard in Sonoita and making his own boutique wine label, Los Milics (see page 160). Perhaps it is the industry's most telling sign that when it comes to Arizona wine, Pavle Milic is here to stay.

SONOITA

A true wild west ranching community, the region of Sonoita is a playground of rolling hills and wavy grasslands. Amid its growing number of wineries and vineyards, Sonoita's offerings are relatively sparse, but its rustic beauty and compelling wine offerings make the region well worth seeking out.

Stay

The Downtown Clifton Hotel

Tucson

downtowntucsonhotel.com

This self-proclaimed Sonoran Boutique Hotel has all the charm of a sleek mid-century-modern hideout with the familiar warm hospitality of a community hangout. It makes for a perfect road-trip stop for anyone on their way between Los Angeles and Marfa, Texas. Sip a few cocktails or a glass of Arizona wine in the Red Light Lounge, or simply enjoy the cozy confines of your room, complete with Netflix-powered television.

La Hacienda de Sonoita

Sonoita

haciendasonoita.com

The four rooms here each have custom iron fixtures made by a local artist, and views of the surrounding landscape, which frequently features grazing deer and antelope.

Next Door @ Dos Cabezas

Sonoita

doscabezas.com/stay-with-us

Now visitors to Dos Cabezas WineWorks can stay in a studio apartment or a two-bedroom house with unique amenities such as an outdoor pizza oven and a record player. Best of all, no designated driver is required, it's right next to the winery and tasting room.

Dine

The Cafe

Sonoita

cafesonoita.com

Owned and operated by local chef Adam Puckle, The Cafe is a small, understated eatery that's big on bringing local produce, meats, and wines to the table. Salads and herbs are sourced from the green garden on the back patio, pasta dishes are soulfully flavorful, and the juicy burgers are best enjoyed with two hands.

Dos Cabezas WineWorks Pizza Truck

Sonoita
doscabezas.com

The Dos Cabezas WineWorks winery has now expanded to include a wood-fired pizza truck. Visitors are invited to sip through a selection of wines and fill up on fresh, house-made pizzas on the colorful outdoor patio.

Feast

Tucson
eatatfeast.com

One of Tucson's most celebrated bistros, Feast offers an upscale yet relaxed setting, featuring a monthly menu that spotlights seasonal local ingredients. Dishes are avant-garde but expertly executed, with items such as rosemary goat cheese risotto, orange-glazed chicken with beets, or house-made ricotta cheese wrapped in beet-cured sea bass and cantaloupe. The wine list is lengthy and extensive, with both international and local selections. Feast also doubles as a wine shop and generously offers every wine on its menu at the retail price plus a $12 corkage fee.

The Steak Out Restaurant & Saloon

Sonoita
azsteakout.com

As the name suggests, this western-themed saloon is just the place for a no-frills steak dinner complete with baked potato and all the fixings. The long-running chophouse also offers a decent bar menu and live country music on weekends.

Time Market

Tucson
timemarket.xyz

This specialty grocery market and bakery offers a wide array of fresh produce and sundry items from local producers. It also serves up pizzas and sandwiches from its counter-service kitchen. (Bonus, the espresso bar is top-notch, as is their on-tap beer selection.)

190   THE WINES OF SOUTHWEST U.S.A.

See

Patagonia Lake State Park

Nogales

azstateparks.com/patagonia-lake/

This 265-acre (107hectare) lake is one of the area's most scenic water-scapes, offering an excellent avenue for fishing, hiking, and general out-door recreation.

Patagonia-Sonoita Creek Preserve

nature.org

Owned by the nature conservancy, this outdoor haven is home to some of the best bird-watching in the state (especially March through September). You'll also find towering Freemont cottonwood trees that are more than 130 years old.

WILLCOX

Another Arizona outpost, Willcox may be where the majority of the grapes used in Arizona wine are grown but don't expect a thriving me-tropolis. Instead, time here slows to a friendly, relaxing pace. Visitors will find the usual pod of fast-food joints and big-brand motels off the freeway, but there are a few other spots worth seeking out for dining and turning in for the night.

Stay

Dos Cabezas Retreat Bed & Breakfast

Willcox

doscabezasretreat.com

This charming property affords spectacular views of the Chiricahua Mountains. The historic guesthouse bedrooms date back to the 1870s and early 1900s. Homemade breakfasts are exceptional, including local-ly roasted coffee, local sausage, fried potatoes, and a home-made Dutch baby pancake with local fruit and prickly pear syrup.

Sunglow Guest Ranch

Pearce

sunglowranch.com

This rustic ranch resort offers the full dude-ranch experience, complete with horseback riding, hayrides, and poolside relaxation with views of the Chiricahua foothills.

Dine

Big Tex BBQ Restaurant

Willcox

bigtexbbqaz.com

Not exactly the real deal you'd expect to find in the Lone Star State, but Big Tex will certainly suffice for those in need of a barbecue fix. Served out of a converted locomotive car across from the town's railroad park, brisket is the way to go, but the burgers are nothing short of delicious. Pair your food with one of several Willcox wines on the menu.

La Unica Restaurant & Taqueria

Willcox

You know you're in the right spot when the locals make a regular stop here. La Unica is the spot for authentic Yucatan-style tacos and standard Mexican fare. The camarones al mojo de ajo are not to be missed.

VERDE VALLEY—COTTONWOOD, CLARKDALE, JEROME, AND CORNVILLE

Just a short half-hour or so from Sedona, the small towns of Cottonwood, Clarkdale, Jerome, and Cornville each represent the Verde Valley. The region has seen a significant uptick in tourism over the past decade thanks in part to the flurry of wineries that have opened in the area. Old Town Cottonwood has a bonanza of delicious dining options, all within walking distance of one another, on Main Street. A 25-minute drive from Sedona, it has become the main stage for the valley's most famous wineries. The historic, hillside mining town of Jerome is a beacon for all who love history, eclectic artistry, wine tasting, and perhaps the best cup of coffee in Verde Valley (you'll find it at the Caduceus Cellars wine tasting room, which doubles as a morning coffee shop). The smaller outposts Clarkdale and Cornville have both seen a revival as well.

Stay

Cottonwood Hotel

Cottonwood
cottonwoodhotel.com

A central spot in the heart of Cottonwood, this boutique inn is housed in an early-twentieth-century building right on Main Street and offers cozy, rustic, homestyle charm.

The Ghost City Inn

Jerome
ghostcityinn.com

Built in the 1890s, this inn began as a boarding house for copper miners in the "Ghost City", aka Jerome. While there are many historical, haunted locales in Jerome, this is a standard-bearer.

Mescal Canyon Retreat

Clarkdale
mescalcanyonretreat.com

Located on the outskirts of Clarkdale, this off-the-grid bed and breakfast is located in an idyllic setting at the foothills of Mescal Canyon. Not only are the rooms comfortable, but also the overall focus on health and wellness throughout the property brings an added level of relaxation.

The Tavern Hotel

Cottonwood
thetavernhotel.com

Converted from a 1925 grocery market, this cozy hotel now houses ten spacious guest rooms, a two-bedroom cottage, and an adjacent bar and restaurant, The Tavern Grille.

The Vineyards

Cornville
thevineyardsbandb.com

This charming, traditional bed and breakfast offers a refreshing break from everyday life. Enjoy a glass of wine on the property's green beneath the shade of surrounding almond trees after touring the region's offerings.

WHERE TO EAT AND STAY IN ARIZONA 193

Dine

The Asylum

Jerome
asylumrestaurant.com

Located inside the historic Jerome Grand Hotel, this former hospital is an excellent stop not only for Southwestern-inspired cuisine but also for its legendary haunted halls. Only hotel guests are afforded a tour of the haunted hotel, but dining here can be just as fun.

Crema Craft Kitchen and Bar

Cottonwood
cremacottonwood.com

A cheerful spot serving up breakfast, lunch, coffee, and gelato with Southwestern flare. Try the smoked turkey and green chile in a demi-baguette, red chile-glazed bacon for breakfast, or local fig gelato for an afternoon snack.

Farside Bistro

Cottonwood
farsidebistro.com

A unique find, this family-owned bistro serves up delicious, authentic Persian cuisine. Tender lamb shank is a consistent crowd-pleaser, but you also can't go wrong with chicken or steak kebabs. (Word to the wise, ask for all the dipping sauces—hummus, shallot, Baba Ghanoush.)

The Flatiron

Jerome
theflatironjerome.com

For a great start to the day, breakfast at The Flatiron is just the ticket. From scrambled egg and toasty egg sandwiches to bagels and seasonal fruit, there's something for everyone at this local favorite.

Merkin Vineyards Tasting Room & Osteria

Cottonwood
merkinvineyardsosteria.com

Although it functions as the Verde Valley tasting room for Merkin Vineyards, you should not step into this place without sitting down for a meal. House-made pastas are complemented by seasonal vegetables, meats, and sauces sourced from local purveyors.

PART 4
COLORADO—HIGH ALTITUDE WINE

13
HISTORY

The thirty-eighth state to join the United States, Colorado is nicknamed the "Centennial State" for joining the U.S. in the year 1876, 100 years after the signing of the nation's Declaration of Independence. Colorado took its name from the Colorado River, or Rio Colorado ("Red River"), as described by early Spanish explorers upon observing the reddish silt of the riverbed.

Major exploration of Colorado began in the late eighteenth century, when French trappers and Spaniards discovered the Rocky Mountains. At the time, the area was already home to several indigenous tribes, including the Nez Percé, the Shoshone, the Crow, the Lakota, and the Ute. In the following century, European conquest would encroach on the tribal lands, with French, Spanish, and American land claims spreading throughout the region.

With the signing of the 1803 Louisiana Purchase, the U.S. government purchased all land east of the Continental Divide from France. Soon after, Meriwether Lewis and William Clark were dispatched to survey nearly 8,000 miles (12,875 kilometers) of the Rocky Mountain West. Their findings over two and a half years encouraged other adventurers and settlers throughout the nineteenth century to move further west, a migration which only accelerated with the completion of the Transcontinental Railroad in the 1860s.

The government signed numerous treaties to defuse tribal objections to the increase in settlement, but ultimately forced tribes onto smaller reservations. Gold and silver mining mania preceded Colorado's entry to statehood in 1876. Mining, grazing, and timber played significant roles in regional economic development, sparking growth in financial and industrial support. Post-Second World War prosperity saw more holidaymakers visiting the national parks and guest ranches, making tourism a leading industry.

Colorado's winemaking history begins more than two centuries after that of the other Southwest states. Grape growing here began as a direct result of the proliferation of fruit growing in the far western part of the state around Grand Junction, where the Gunnison River flows into the Colorado River on its way southwest. This sunnier, warmer part of the state is home to protective canyons along these rivers that shelter crops against the prevailing cold and wind of the continental region. Although peaches, pears, plums, apricots, and cherries were the first fruit plantings, documentation from an 1895 account called *The Fruit Belt of Mesa County Western Colorado* reveals that grapes such as "Black Hamburg, Flame Tokay, Zinfandel, Sultans, Muscat, and Malaga grow to perfection in the open, and this is the only valley in the State where it is possible to grow them at all."

According to this publication, settlement into Mesa County in the western part of the state began in the early 1880s after the native Utes were forced to relocate to a reservation in Utah. With the arrival of the railroad in 1882, which extended a path from the east through the Rocky Mountains and on through the Salt Lake of Utah, settlers were quick to take advantage of the region's fertile soils and ample water resources. To utilise the nearby Colorado River, gravity canals were constructed to divert water into the mouth of DeBeque Canyon near Palisade. In 1890, Colorado Governor George A. Crawford, who founded Grand Junction in 1881, planted 60 acres (24.3 hectares) of grapevines and other fruit near Palisade. While the region would soon develop into the state's veritable fruit basket, today producing some of the country's most celebrated peaches and cherries, these early grape plantings became a foundational part of Colorado's wine-growing future.

By 1909, the U.S. Department of Commerce Census of the United States reported that Colorado had 1,034 farms involved with grape production, with more than 254,000 vines of bearing age and more than 101,000 vines of pre-bearing age. But, as with other individual states leading up to the 1920s, the General Assembly of Colorado enacted its own statewide Prohibition, a full four years before the passage of the Eighteenth Amendment.

Colorado mining history

In 1848, prospector James W. Marshall found gold at Sutter's Mill in Coloma, California. News of his discovery spread swiftly to the eastern United States, igniting the historic California Gold Rush, which would see more than 300,000 people brave the yet undeveloped western frontier to journey to California's sunny landscape in search of great fortune. But California wasn't the only gold-rich state. Mines began popping up for the precious ore all over the country, including in Colorado. In fact, the early 1850s would see several parties of prospectors stopping short of California to pan for gold in the streams of the Rocky Mountain foothills.

It wasn't until 1858 that the first significant gold discovery took place near Denver, at the mouth of the Little Dry Creek, setting off the Pikes Peak Gold Rush (later known as the Colorado Gold Rush). In the years that followed, a flurry of roughly 100,000 gold seekers staked their claim in Colorado, setting up mining camps further west into the Rockies in Idaho Springs, Breckenridge, South Park, Leadville, Summitville, Telluride, and Cripple Creek. Though gold wasn't always found in abundance, miners soon found a plentiful supply of silver, copper, and lead. (Later, uranium would also be found in abundance, making Colorado the third-largest uranium reserve in the U.S., behind Wyoming and New Mexico.)

In Colorado, gold production was estimated at 270,000 ounces (7.7 million grams) in 1892. By 1900, it had peaked at about 1.4 million ounces (39.7 million grams). Overall production slowly decreased throughout the twentieth century. Today, there is only one gold mine in operation, the Cripple Creek & Victor Gold Mine near Colorado Springs, which yielded about 320,000 ounces (9.1 million grams) of gold in 2019.

Although seemingly great wealth was drawn from Colorado's mining era, the ecological repercussions have been a most significant drawback. The Colorado Geological Survey conducted a study between 1991 and 1999 of the primary watersheds affected by the mines (there are more than 18,380 abandoned mines in the state). Of the mines analyzed in the survey, 917 showed levels of significant or extreme degradation and mineral hazards. The damage is caused by acid mine drainage, which results when surface and groundwater combine with the sulfuric rock dredged up from the mines, to create sulfuric acid. The acid causes leaching of iron, manganese, lead, copper, zinc, and cadmium from the rocks, and drains into surface water. While many efforts have

200 THE WINES OF SOUTHWEST U.S.A.

> been made to remediate mining sites, those that are abandoned no longer
> have an entity accountable for the costs associated with the clean-up.
>
> Acid mine drainage is said to have affected 89 percent of Colorado streams,
> resulting in environmental concerns for wildlife, fish, and surrounding vegeta-
> tion. While regional efforts are in place to reduce water contamination, the
> process is long and expensive. It is estimated that the Colorado Department
> of Public Health and Environment treats nearly 98 million gallons (370 million
> liters) of water a year from a total of five mining sites.

As with the rest of the country, all wine production was halted during Prohibition. According to Thomas Pinney, in his book *A History of
Wine in America: From Prohibition to the Present*, the growth of wineries in operation after the 1933 repeal of Prohibition was a slow process.
Part of this was due to the need for more vineyard plantings, but it was
also down to new regulations which had been put in place that extended
well beyond producing wine as part of a simple farm. In 1937, the total
number of wineries in America was estimated at 1,206. California was
home to the lion's share with 630 wineries. Ohio came in second with
130 wineries, and New York had 123. At the time, Colorado had only
two.

Many credit Dr. Gerald Ivancie, a Denver dentist, with launching the
modern Colorado wine industry in 1968, with Ivancie Cellars in the
state capital. Although it was one of the state's first wineries of the modern wine era, Ivancie was merely making wine from California grapes
rather than Colorado wines. In 1972, Ivancie's hobby business was sold
to new owners who aspired to make authentic Colorado wine, partnering with a few small growers in western Colorado with experimental
Vitis vinifera plantings. Sadly, the operation was short-lived, closing its
doors in 1975.

It seems likely that the Four Corners Regional Commission, which
had been launched through federal legislation to research the feasibility of grape growing in the region, helped foster further aspirations for
wine production in Colorado. By 1974, an experimental viticultural
vineyard was planted by Colorado State University in Orchard Mesa,
near Grand Junction. As this vineyard was coming into production,
in 1977, the state of Colorado passed a farm winery law to encourage
more vineyard plantings.

One of the first to take the plunge was Colorado Mountain Vineyards, which opened its doors in 1978 in Golden, 16 miles (25.5 kilometers) west of Denver. Soon after, in 1984, Pike's Peak Vineyards opened near Colorado Springs, and in 1985 Plum Creek Cellars was established in Larkspur. As the vineyards in Palisade began to take root, both Colorado Springs and Plum Creek relocated to Palisade. Simultaneously, in 1982, the handful of producers in the industry organized the Rocky Mountain Association of Vintners and Viticulturists, which in 1990 evolved into the state-funded Colorado Wine Industry Development Board.

For the better part of the 1990s and early 2000s, Colorado wine could broadly be described as "finding its way", and was driven more by tourism than by quality. Nevertheless, the wine industry has steadily grown at an average pace of about 13 percent each year since 1992.

Today, with more than 1,000 acres (404.7 hectares) of vineyard plantings and around 140 licensed wineries (which includes a handful of cider makers and meaderies), the state produces upwards of 150,000 cases annually. In recent years the relatively young industry has been aptly described as "becoming".

14

CLIMATE, REGIONS, AND CHALLENGES

With its soaring mountain peaks, rushing river rapids, thick Aspen forests of whitewashed bark and fluttering leaves, and expansive mesa vistas, Colorado is a year-round outdoor playground for thrill-seekers and nature lovers alike. From winter skiing and summer trekking among the state's snow-capped peaks to white-water rafting and fly-fishing its 6,000 miles (9,656 kilometers) of fishable streams, many have sought adventure and solace in the state's dramatic landscapes. Colorado's defining geographical feature is the majestic chain of mountains, the southern Rocky Mountains, whose jagged peaks bisect the state. Its eastern border is at the western edge of the American Great Plains.

The state is notable as the only U.S. state to lie entirely above 3,281 feet (1,000 meters) in elevation. Its capital city, Denver, is nicknamed the "Mile High City" due to its elevation above 5,200 feet (1,585 meters). The lowest point of the state is in Yuma County, in the northeast, which sits at 3,300 feet (1,005 meters). As the eighth most extensive state in the U.S., Colorado's geography is as diverse as it is dramatic, boasting alpine mountains, high plains, sand dunes, deep canyons, and rushing rivers. Home to four National Parks (Rocky Mountain, Mesa Verde, Great Sand Dunes and Black Canyon of the Gunnison), eleven National Forests, and two National Grasslands, Colorado lies astride the highest mountains of the Continental Divide. Four of the nation's major rivers have their source in Colorado: the Colorado, the Rio Grande, the Arkansas, and the Platte. Additionally, all of the state's rivers originate within its borders and flow outward, except for the Green River, which flows across its northwestern corner.

CLIMATE

Not surprisingly, Colorado's climate is as diverse as its geography. One might assume that the southern part of the state would be warmer than the north, but given the combinations of mountains, foothills, high plains, and desert lands, this is not always the case.

In general, this diversity in geography results in a cool, dry, refreshing climate with wide seasonal temperature swings and broad diurnal shifts from day to day. Summer ushers in hot days in the plains, quelled by afternoon thunderstorms. In contrast, the mountain regions are nearly always cool. Statewide, humidity is generally low, making temperatures feel bearable, even during hot days. Higher elevations throughout the state result in thinner atmosphere bringing greater penetration of solar radiation. At night, temperatures drop quickly, and freezing temperatures are not uncommon in mountainous locations year-round. Differences in elevation significantly affect temperatures in various localities.

Colorado's annual precipitation statewide is about 17 inches (432 millimeters) but ranges from 7 inches (178 millimeters) in the middle of south-central Colorado to more than 60 inches (1,524 millimeters) in a few mountain locations. Temperature swings and precipitation generally increase with altitude, but are modified by the orientation of mountain slopes with respect to the prevailing winds.

In general, precipitation tends to be less in lower elevations as a result of the state's distance from the Pacific Ocean and the Gulf of Mexico. Eastward-moving storms from the Pacific Ocean lose much of their moisture as rain or snow on the mountaintops and westward-facing slopes, leaving little to fall on the eastern slopes. Southward-moving storms bring polar air that results in sudden drops in temperature, par-ticularly in the vast and expansive plains sections of the state. This cold air is frequently too shallow to cross the mountains to the western por-tion of Colorado, resulting in milder weather.

In other instances, the plains experience "chinook winds", brought on when the region is covered with a shallow layer of cold air and hit suddenly by strong westerly winds. Warmed by rapid descent from higher levels, these winds bring significant and sudden temperature rises of 25–35°F (14–19°C) within a short period of time. The eastern part of Colorado receives the most torrential rainfall from April through early September, while southern and western Colorado tend to receive moisture from mid-July into September, as part of what is widely referred to as the "Southwest Monsoon".

Vineyards in Colorado are mostly nestled in the temperate, high elevation river valleys and mesas of Mesa and Delta counties, with some acreage in Montezuma County in the southwest. They face hot days followed by cool nights, with typical wide diurnal temperature variations.

The great divide of the Rocky Mountains

One of the defining features of the American Southwest is the southern chain of the rugged Rocky Mountains. This majestic chain, which stretches for more than 3,000 miles (4,828 kilometers) from its northern tip in Alaska to its southern edge in western New Mexico, has lured myriad outdoor adventurers and nature lovers to heed its perpetual call to the wild. Mountain chains formed between 80 million and 55 million years ago as a result of tectonic forces. During this time, plates began sliding underneath the North American plate, which resulted in the formation of a significant mountain range running down western North America. The dramatic peaks and valleys of the Rockies continued to be defined with further tectonic activity and glacial erosion.

Of the 100 highest peaks in the Rockies, 78 are located in Colorado, including the 30 highest. The state's majestic landscape boasts natural hot springs, hundreds of lakes and rivers, twelve National Parks and Monuments, and 58 mountain peaks over 14,000-feet (4,267 meters), more than any other state. The highest peak is Mount Elbert at 14,433 feet (4,399 meters).

Commanding the eastern borders of Idaho and the western boundary of Montana, almost the whole of Wyoming, and just dipping into north-central New Mexico, the Rocky Mountains virtually divide the rectangular state of Colorado down the middle. This unique bisection through Colorado's center defines part of the Continental Divide, an imaginary line along the high points of the Rocky Mountains that splits the continent, directing the natural flow of water either east to the Atlantic Ocean, or west to the Pacific. Colorado is where this divide, also commonly known as the Great Divide, reaches its highest point at the summit of Grays Peak, whose elevation is 14,278 feet (4,352 meters). Hiking the Colorado portion of the Continental Divide Trail, which threads from its northernmost boundary to its southern tip, has become a favorite pastime for outdoor enthusiasts as much of it remains above the treeline, providing miles of unobstructed views along its route.

GEOLOGY

The geology of Colorado is the result of island arcs that slowly accumulated onto the edge of the Wyoming Craton, which is considered part of the initial core of the continental crust of North America from more than 1.8 billion years ago. The land was primarily a shallow sea before the Rocky Mountains uplifted from the Sonoma orogeny more than 300 million years ago with intense volcanic activity. This was followed by the Laramide orogeny from more than 70 million years ago. Erosion over the next 20 million years cut down the mountain bedrock. At the same time, a series of other volcanic uplifts produced the San Juan Mountain range at the southernmost part of the Rocky Mountain chain. What followed during the Miocene era was a large-scale, region-wide uplift that raised the entirety of Colorado, New Mexico, Utah, and Arizona 5,000 feet (1,524) above sea level. Along the ridges of the Rocky Mountains chain is the Continental Divide of the Americas, which designates the line where waters either flow westward towards the Pacific Ocean or eastward towards the Atlantic Ocean.

With an average altitude of about 6,800 feet (2,073 meters) above sea level, Colorado is the highest contiguous state in the Union. Roughly three-quarters of the nation's land above 10,000 feet (3,048 meters) altitude lies within its borders. The state has 58 mountains that tower at 14,000 feet (4,267 meters) or higher and about 830 peaks that range between 11,000 and 14,000 feet (3,352–4,267 meters).

REGIONS

Colorado has two official AVAs: the Grand Valley and the West Elks. Though there are noteworthy plantings in other parts of the state, including the southwest corner near Cortez, more than 90 percent of the state's grapes are grown in these two main regions.

The Grand Valley

Located on the Western Slopes of the Rocky Mountains, between the towns of Palisade and Grand Junction, the Grand Valley is the heart of Colorado grape growing. Historically famous for growing some of the best orchard fruit in the country, including Palisade peaches, plums, and cherries, the Grand Valley has also grown grapes for more than a century. With nearly 800 acres (323.7 hectares) under vine, the region, which received its AVA designation in 1991, accounts for more than 80 percent of Colorado's overall plantings.

Situated along the Colorado River, about 40 miles (64 kilometers) east of the Utah border, the region is defined at the mouth of the DeBeque Canyon to the eastern part of the region. As it opens westward along the river, it reveals a lush, green valley floor at the town of Palisade. The area is uniquely sheltered by the largest flat-top mountain in the world, Grand Mesa. Further west, the region spills into the East Orchard Mesa and Orchard Mesa at the south bank of the river and continues west of Grand Junction to the base of the Colorado National Monument.

Vineyard elevations range between 4,000 and 4,800 feet (1,219–1,463 meters) on an east–west series of chalky mountains called the Book Cliffs, which face south. This mountain chain reflects solar energy onto the valley floor, offering consistent warmth through the region. Breezes from the Colorado River keep air flowing through the eastern vineyards, located closer to the mouth of DeBeque Canyon, cooling vineyards during the hot summers and providing warmth during cold winters.

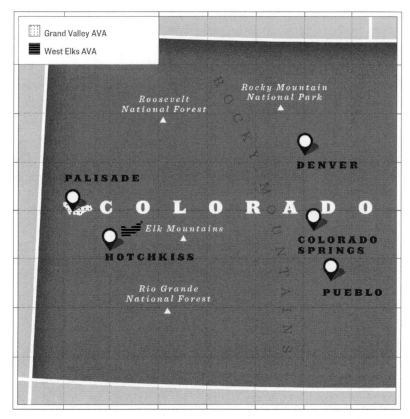

208 THE WINES OF SOUTHWEST U.S.A.

Most of the vineyards are located in the eastern portion of the region as westward sites are more susceptible to temperature extremes. The average annual rainfall for the region is 8–9 inches (203–229 millimeters), making irrigation a necessity. The soils are primarily eroded sediment washed down from the surrounding mountains and mixed with a prehistoric sea bed. In some areas, this sediment includes river cobble from the Colorado River. Other areas have remnants of basalt from the washout of the Grand Mesa, while parts of the region include sandstone from the Book Cliffs. Each of these three different rock types is well-draining and predominantly alkaline (7.2–8.5 pH).

At the peak of summer, temperatures can reach as high as 93°F (34°C) but dip down more than 30 degrees Fahrenheit (16.6°C) to the low sixties (the mid-teens Celsius) in the evening. In general, this wide swing in temperatures makes for ideal growing conditions. The region's primary challenges are late spring and early fall frosts. But perhaps more significant are the nearly 300 days of intense sun exposure, which produce considerable vigor in the vines. As a result, growers have to be vigilant with canopy and fruit management throughout the growing season.

Along the warmer valley floor, producers work with Rhône varieties such as Syrah, Viognier, Grenache, Mourvèdre, and Roussanne. But in the cooler higher elevation sites, varieties such as Merlot, Cabernet Franc, Malbec, and Cabernet Sauvignon are more prominent.

West Elks AVA

About 65 miles (104.5 kilometers) southeast of the Grand Valley, between Hotchkiss and Paonia along the North Fork of the Gunnison River, is the West Elks AVA. Significantly smaller than the Grand Valley, West Elks accounts for about 10 percent of Colorado grape production but received its AVA designation in 2000. Small vineyard parcels, often averaging about 3–5 acres (1.2–2 hectares), are scattered among the farming communities of this region at the foothills of the West Elks mountain range.

Elevations here range between 5,500 and 7,000 feet (1,676–2,134 meters), making it one of the highest wine-growing regions in the world outside of Argentina. The rocky soils of the region are distinctive, particularly on the north side of the Gunnison River. Volcanic, basalt-based clay topsoils that have washed out over the valley from the Grand Mesa overlay a sedimentary layer with caliche and unconsolidated alluvial soils. While it generally takes new vineyard plantings a few years to establish, while the roots push down beneath the topsoils, older vineyards which have reached the calcium-rich caliche layer are said to yield wines with more minerality and nuance.

The average rainfall in West Elks is slightly higher than in the Grand Valley, with about 12.5 inches (318 millimeters) annually. Due to its higher elevation, the growing season generally begins about two weeks later, and also ends earlier, with about 30 percent fewer days between the last spring frost and the first fall frost. This shorter season offers growers a more limited selection of varieties to plant. Cool-climate grapes such as Pinot Noir, Riesling, Pinot Gris, and Gewürztraminer are among the region's most common plantings.

Other regions

Beyond the two official AVAs, it's worth noting that some of the state's promising wine producers are growing grapes in outlier regions such as in south-central Colorado near Canyon City, east of the Rockies near Denver, Boulder, and Colorado Springs, and in the far southwest part of the state near Cortez and Montezuma in the Four Corners region.

The Four Corners region

In the southwest corner of the state, around McElmo Canyon, the Montezuma Canyon, and Mesa Verde National Park, a few outlier vineyards have been planted. This raw, desert landscape is vast and arid, peppered with scrub brush and ancient Puebloan ruins. In stark contrast to the growing conditions of the Grand Valley and West Elks, this area is most similar to vineyard sites found in New Mexico and Arizona. Vineyards here are usually located above 5,000 feet (1,524 meters), with one outlier vineyard of Riesling at just under 7,000 feet (2,134 meters). Most of the soils are sandy-loam clay that is well-draining, with zero organic matter and high pH. In fact, growers often have to amend soils with the use of microbes and compost to help with nutrient uptake. Summer monsoons are fairly common, threatening rot and humidity at the peak of the harvest season. This landmark Four Corners region at one point joins Colorado, Utah, Arizona and New Mexico.

The Front Range

While the majority of Colorado vineyards are planted in the western part of the state, it's important to note that the majority of tourism for

210 THE WINES OF SOUTHWEST U.S.A.

Colorado wine is east of the Continental Divide around cities such as Denver, Boulder, and Colorado Springs. In fact, more than half the state's wineries are located here, forcing growers to truck grapes more than 200 miles (322 kilometers) over the Rocky Mountains during harvest. Growing conditions in the Front Range are significantly different from those on the Western Slope. Temperatures average about 10 Fahrenheit degrees colder (roughly 5.5 Celsius degrees) during the winter and there are also wider swings on any given day. As a result, *Vitis vinifera* does not do well here, but some growers are experimenting with native grape varieties such as Norton as well as French hybrids such as Chambourcin and Seyval Blanc, all of which are more cold hardy.

GRAPES

Colorado offers a generally cooler climate in parts of the state than its neighboring state to the south, so Bordeaux and German varieties have had much greater success. Cabernet Sauvignon accounts for 14 percent of overall vinifera plantings, followed by Merlot at 12.8 percent, and Cabernet Franc at about 9 percent. With more alkaline-rich soils throughout its growing regions, the Bordeaux varieties tend to compare more to the Old World in terms of style, with more earth and other tertiary characteristics than is commonly found in Californian examples. Riesling accounts for about 9.3 percent of overall vinifera plantings, and can mainly be found in the West Elks AVA. With more than 300 days of sunshine in the region, Riesling tends to take on riper, Alsatian characteristics and is often produced in a dry style, though sweet offerings can also be found. West Elks is also home to Pinot Noir, which accounts for nearly 2 percent of overall *vinifera* plantings. Gewürztraminer makes up just over 3 percent, and Pinot Gris about 1.5 percent. In the warmer regions of the Grand Valley AVA and in the southwest corner of the state, known as the Four Corners region, Rhône varieties have grown in popularity. In these regions, Syrah (5.8 percent) and Viognier (3.4 percent) lead Rhône plantings, though Grenache, Cinsault, Roussanne, and Mourvèdre are also gaining ground.

The main issue when growing grapes in Colorado is cold weather events. On average, significant vascular damage to a vine begins when temperatures fall to between 2 and 4°F (-16° and -15°C). Vine kill occurs at -15°F (-26°C). A few hard winters in a row can cause a grape grower to consider numerous alternatives. In recent years, the idea of cold-hardy hybrid grapes has taken hold with grapes such as Frontenac, Chambourcin, St. Vincent, Aramella, and La Crescent showing promise. By comparison, many of these hybrid varieties can withstand temperatures that dip as low as -35°F (-37°C). Today, non-*vinifera* grapes account for about 18 percent of overall vineyard plantings in the state.

CHALLENGES

What growers have learned to be true of viticulture in Colorado is that it can be as rugged, extreme, and forbidding as its dramatic landscapes. One of Colorado's most significant grape-growing challenges is the relatively short growing season. Compared to warmer regions further south, it can be two full weeks shorter at either end of the season. While grapes need enough time to ripen, it's also crucial for the vine's growing cycle to have enough time to harden off after harvest before the winter sets in.

For instance, 2018 and 2019 yielded bumper crops for growers in both the Grand Valley and West Elks. In fact, the supply of grapes outstripped demand in both vintages. In 2018 alone, nearly 400 tons (363 metric tons) of grapes were left on the vines. But in the years prior, fall frosts decimated vineyards. In 2014, many growers lost up to 75 percent of their plantings, having to cut vines down to the ground and start again. The broad spectrum of circumstances has made it difficult to gauge the best growth strategy for producers. Planting more acreage either means more potential frost damage or having more fruit than the market demands. One way growers have mitigated this is by launching their own wine brands to diversify their businesses and having landing places for fruit in abundant years.

While trying to plant in warmer, more buffered sites has been a longtime goal, using cold-hardy grape varieties such as Chambourcin, Frontenac, and Marquette, which can withstand dramatic shifts in temperature, has become a growing trend among producers. These hybrids also help in the fight against phylloxera, which has already affected more than 100 acres (40.5 hectares) of vineyards in the past few years.

But while hybrid grapes may grow well and make decent wines at friendly prices, their status as lesser-known varieties makes them difficult to sell to the average consumer. In many regards, it's the general consumer that remains one of the state's greatest challenges. Recent years have seen more and more retailers and restaurants embrace quality wines from some of the state's best producers. But by and large, local consumers remain skeptical, a fact that will no doubt change with time, as the quality of Colorado wine continues to increase.

Because the region is unlike any other in the world, there is no set template. As such, using the power of a strong community of producers is a critical factor in pushing the industry forward. The more transparent about their successes growers and winemakers are, the more they will learn and the better they will become at what they do.

15

COLORADO PRODUCERS

GRAND VALLEY AVA

Carlson Vineyards
Grand Junction
carlsonvineyards.com

A Colorado wine pioneer, Carlson Vineyards evolved in 1981 out of a small vineyard planting in an old apricot orchard by Parker and Mary Carlson. At first, they made wines with the grapes as a hobby, but soon developed a passion for the endeavor, prompting them to open a winery in 1988, which they housed in a 1930s fruit packing shed on four acres (1.6 hectares) on East Orchard Mesa just a few miles south of Palisade. Since then, the small farm winery, which is also one of the region's significant orchards, has earned a reputation for producing premium Colorado wines from both grapes and other Palisade-area fruit. Early on, the Carlsons became well-known for Riesling and Gewürztraminer (both dry and sweet), as well as Merlot, Shiraz, Chardonnay, and Lemberger. For nearly three decades, the winery steadily grew to expand into a 4,000 square-foot (372 square meters) facility and now produces about 10,000 cases annually.

In 2015, the winery changed hands when Garrett Portra, who had been working as an assistant winemaker there since 2011, purchased it from Carlson. Carlson Vineyards has maintained its fan-favorite line of off-dry wines and fruit wines, but Portra has pivoted to also include bigger, more full-bodied reds and dry white wines. While both the Riesling and Gewürztraminer are classics at Carlson Vineyards, the peppery, red-fruit dominant Tyrannosaurus Red (T-Red), made from the German

214 THE WINES OF SOUTHWEST U.S.A.

grape variety Lemberger, is a consistent pleaser. Of the fruit wines, be sure to give the plum wine a try.

Colorado Cellars

Palisade

coloradocellars.com

Colorado Cellars claims the distinction of being the first winery in the post-Prohibition era to make and sell wine from Colorado-grown grapes. Established in 1978 by a group of like-minded investors, the winery was a pioneer for the modern Colorado wine industry. The wines were initially sold under the Colorado Mountain Vineyards brand with the Colorado Cellars brand appearing in 1986 and the Rocky Mountain Vineyards and Orchard Mesa Wine Company brands appearing in 1990. In 1989, Richard and Padte Turley took over full ownership of the winery from the last of the twelve original investors. In 1991, they helped to establish the Grand Valley American Viticultural Area.

Nestled in a hillside overlooking the entire Grand Valley, the winery is home to the last of the original test plot vineyards of the 1974 Four Corners Project. In the vineyard, the Turleys practice natural growing techniques: leaf removal (thinning) and canopy management, the re-introduction of grape pomace as fertilizer, minimal natural pest control, measured irrigation application, and gentle hand harvesting. Padte Turley holds the distinction of being Colorado's first female winemaker. In the cellar, she employs extensive cold fermentation with native yeasts, natural fining and filtration, exclusive use of stainless steel fermentation tanks, and a barrel program which includes American, French, and Hungarian oak. Over the years, the brand has grown to include more than 30 types of wines (including traditional, dessert, sparkling, and fruit wines), with an annual production of about 12,000 cases. Though they have garnered a reputation for excellent fruit wine, the Cabernet Sauvignon and Merlot are consistent crowd pleasers, as is their luscious Port-style wine.

Colterris

Palisade

colterris.com

The Colterris story is essentially the fulfillment of a promise. When Scott High proposed to his wife, Theresa, he promised her a vineyard. At the time, he was the owner of a wine distribution company in Denver and Theresa was in wine sales. Although the vineyard would take some time to acquire, the two made their first move towards fruit production with the purchase of a 10-acre (4-hectare) peach orchard in Palisade in 1999. Over the years, that peach orchard would grow to include more than 70 acres (28.3 hectares) of land, with more than 30,000 peach trees in production under the family-made brand, High Country Orchards. (Natural and organic supermarket Whole Foods Market has been the exclusive distributor of High Country's harvest for more than a decade.)

But the Highs' prolific success with peach farming didn't overshadow their original intent to grow grapes; it merely helped to finance that intention. In 2006, they planted 20 acres (8.1 hectares) of Cabernet Sauvignon. A few years later, in 2010, they released their first vintage under the name "Colterris". The name was an invention of their oldest son, who combined the "Col" of Colorado with *terris*, which is Latin for "from the land".

The venture would only grow from there. Today, the family has expanded its vineyard holdings to include more than 180 acres (72.8 hectares) spread among seven different sites, from the mouth of DeBeque Canyon to East Orchard Mesa, a few miles south of Palisade. Colterris is one of the largest and most technologically advanced wine producers in the Grand Valley, making wines from 100 percent Colorado-grown grapes.

In 2016, the Highs hired Bo Felton to take over as head winemaker. Felton had previously made wine with Duckhorn Wine Company's Goldeneye and Migration Wine division as well as in New Zealand at Rapaura Vintners. Felton was joined by Assistant Winemaker Justin Jannusch, who had spent the previous decade making wines for BookCliff Vineyards in Boulder (see p. 222). Together, the two have strengthened the Colterris brand, putting forth wines that are both powerful and finessed, and expressive of the unique growing conditions of the high elevation Grand Valley. Colterris excels with Bordeaux variety wines. Cabernet Sauvignon, Merlot, and Cabernet Franc consistently make exceptional wines. Also of note is the Coral White Cabernet Sauvignon, a signature offering from Colterris that presents more like a crisp, light, berry-rich rosé with its faint rosy hue.

Plum Creek Winery

Palisade
plumcreekwinery.com

As two of the pioneers of Colorado's modern wine industry, Doug and Sue Phillips have a vision of making quality wines from Colorado fruit. In 1980, they planted their first vineyard on 11 acres (4.5 hectares) of leased property in Paonia. Opting for cool-climate varieties, the Phillipses planted Riesling, Gewürztraminer, and Cabernet Franc. In 1984, they launched Plum Creek Winery in Palisade, where they also began planting vineyards. Today, the Phillipses farm 65 acres (26.3 hectares) of vines between the Grand Valley and West Elks sites. Now, more than 40 years after its inception, Plum Creek Cellars produces about 13,000 cases of wine annually. Winemaker Jenne Baldwin-Eaton specializes in producing wines with an Old-World style that are energetic and beautifully balanced. Principal varieties include Chardonnay, Merlot, Riesling, and Cabernet Sauvignon.

Red Fox Cellars

Palisade
redfoxcellars.com

Unconventional; avant-garde; unorthodox; experimental. All these words fittingly describe the work that Red Fox Cellars has been doing since it opened its doors in 2014. Owners Scott and Sherrie Hamilton, who named the company for the red foxes that scamper about the region, approach their winemaking program respecting traditional techniques, while also experimenting with new ideas. Their son, Kyle, is also part of the team. While straightforward Cabernet Sauvignon, Tempranillo, and Barbera are all well made and delicious, Red Fox also offers Merlot aged in used bourbon barrels, and Cabernet Franc aged in rye barrels. It's all done with the intention of making serious wine, without taking themselves too seriously. These wines are bold, playful, and certainly unique.

The winery's unpretentious tasting room offers an atmosphere more like that of a craft brewery pub. Try the Freestone Peach, a peach wine aged in rye whiskey barrels and the Loca-Motiv, a Chardonnay aged in tequila barrels. The more traditional Teraldego is among the best you'll find in the country. They also serve wine cocktails and cider and fruit wine flights.

Phylloxera in Colorado vineyards

The greatest threat to *vitis vinifera* vines planted on their own roots is the root louse or aphid *Phylloxera vastatrix*. While many Colorado growers were aware of the potential risk, the general understanding was that it was better to plant own-rooted vines to avoid exposing the graft zone of young plants to the winter cold, risking the survival of the entire vine. The high proportion of sand in the soils was also thought to be a sufficient deterrent. As a result, the vast majority of vineyards in both Colorado AVAs were own-rooted. Surveys found no evidence of phylloxera in Colorado's commercial vineyards until the fall of 2016.

The discovery sparked a flurry of concern among Colorado's top grape growers. Although it is impossible to treat vines against the root louse, growers have adjusted vineyard practices, thoroughly cleaning all harvesting and cultivating equipment, and taking care to isolate clothing and footwear to individual vineyard sites. Watching carefully for signs of yellowed leaves, stunted growth, and other symptoms that mimic nutritional deficiencies has helped. The one positive for vineyards infected with phylloxera is that the cold weather throughout the state has prevented a rapid spread. Growers can manage a few vintages with a decent yield before the vines begin to struggle significantly. To counteract the slow decline, many growers have begun planting new grafted vines in between the infected ones. The idea is that the infested vines will still be producing even as their yields are declining, giving the new plantings the necessary three to five years to establish and begin producing viable fruit. Others have committed to destroying vines as soon as they test positive for the infestation, planting new, grafted vines in their place.

Sauvage Spectrum

Palisade

When you're named after a geologic formation from the Grand Canyon National Park, there's a likelihood your future will somehow include working in the soil. Such is the case with Kaibab Sauvage. (The name Kaibab refers to the Kaibab Limestone, the uppermost layer of rock at the Grand Canyon.)

Born and raised on the Western Slope, Sauvage grew up on a five-acre (two-hectare) peach farm. Following his college studies in biology and

218 THE WINES OF SOUTHWEST U.S.A.

chemistry, he turned his attention to grapes, landing a job as a vineyard manager with a local grower. Leveraging his relationships with other peach farmers and grape growers in the region, Sauvage brokered land leases for planting his own vineyards to sell fruit to wineries. After 20 years of building his business, Sauvage now farms more than 60 acres (24.3 hectares) of vineyards and sells to more than two dozen wineries. He also consults for a few private vineyards.

After the 2018 and 2019 vintages, Sauvage was faced with an over-abundance of fruit – the bountiful harvest had him scrambling to make new decisions for windfall crops in good vintages. In 2019, he partnered with winemaker Patric Matysiewski to launch the Sauvage Spectrum label. With an overall focus to deliver fun, fresh and fruit-forward wines, Sauvage Spectrum released three sparkling offerings (dry, off-dry, and *pétillant naturel*) called Sparklet. In the future, Sauvage also has plans to develop still wines.

WEST ELKS AVA

Alfred Eames Cellars
Paonia
alfredeamescellars.com

Alfred Eames Petersen has been making wine since the 1970s. Petersen, who goes by "Eames", is from Wisconsin, and began his wine journey as a hobbyist who would experiment with just about anything he could find, including dandelions. It was his travels to Spain in the early 1960s that piqued his curiosity: enamored of the great wines of Rioja, he began to consider making the hobby into a full-time job. It wouldn't be until 1984 that he would settle on doing so in the remote West Elks region, in Paonia. He planted a small vineyard, Puesta del Sol Vineyards, and in 1994 established the eponymous Alfred Eames Cellars.

Petersen grows exclusively Pinot Noir, the only red grape he thinks suited to the region (he's not particularly interested in white wines). He also sources bigger reds from the Grand Valley, making bold, robust offerings of Tempranillo, Syrah, and Bordeaux blends. The wines of Alfred Eames Cellars are beautifully fresh and lively; the Pinot Noirs are akin to fruity, yet elegantly structured Sta. Rita Hills or Sonoma offerings.

Petersen has now handed over much of the operation to his son, Devin Eames Petersen, who continues his father's tradition of making genuinely sophisticated red wines. A visit to Alfred Eames Cellars

COLORADO PRODUCERS 219

requires an appointment; they do not have a tasting room open to the public. Instead, visits are with Devin or Eames and include a glimpse of the underground cellar, which feels a bit like stepping into the Old World.

Jack Rabbit Hill Farm

Hotchkiss
jackrabbithill.com

This family-owned working farm situated at an elevation just above 6,000 feet (1,829 meters), in Hotchkiss, is Colorado's only certified biodynamic winery. In 2000, Lance Hanson left the software industry in California for the mountain life of Colorado, and began planting grapes on his 70-acre (28.3-hectare) property. From the beginning, he used organic methods to grow the best possible fruit he could in the challenging mountain conditions. But following inspiration from other regions of the world, in 2006 he began transitioning to biodynamic farming. Today, Hanson grows a combined 18 acres (7.3 hectares) of grapes, medicinal herbs, and pasture for sheep and cows.

Hanson and his wife, Anna, work the farm and make the wines themselves. As an extension of his farming methods, the couple employs a minimal approach in the cellar, yielding wines that are fresh, lively, and expressive. Native yeast fermentations are typically complete within 5–6 months. White wines are aged on the lees in stainless steel vats. Reds are aged in neutral French oak barrels for 18–20 months. Try the delicately floral Gewürztraminer, with notes of Palisade peaches and rosehips, and the vibrant, golden-toned Riesling. In addition to wine, Jack Rabbit Hill Farm produces eau de vie, brandy, cider, gin, and vodka.

Leroux Creek Vineyards and Inn

Hotchkiss
lerouxcreekvineyards.com

Leroux Creek is both a winery and guest inn nestled within the scenic landscape of the West Elks AVA. Upon visiting the region in 1999, husband and wife Yvon Gros and Joanna Reckert were reminded of Gros's native home in southern France. So strong was the impression that the pair opted to leave their existing home in the mountain ski village of Vail and exchange it for a more pastoral setting. They purchased a 54-acre (21.9-hectare) property and began work on a vineyard and boutique hotel. The couple has specialized in growing cold-hardy French and American hybrids, including Cayuga and Chambourcin. Due to the high elevation and low disease pressure of the West Elks AVA, the Leroux vineyards are farmed using organic practices. The Chambourcin makes a rich, jammy red with hefty structure as well as a refreshing fruit-forward rosé. Also worth a taste is the My Cherry Amour, a cherry-based wine with a tart, fruity flavor, and notes of sweet almond.

Stone Cottage Cellars

Paonia
stonecottagecellars.com

Stone Cottage Cellars is the realization of the joint dream of Brent and Karen Helleckson to raise a family on a country farm—both are from family farms; he from Minnesota and she from North Dakota. In 1992, after falling in love with the region during a weekend getaway from the Front Range, the Hellecksons left Boulder and their respective careers in aerospace engineering and software. The two began their search for land near Paonia, and when an idyllic parcel, already planted with a few vines, became available, the longtime hobbyist winemakers seized the opportunity.

The Stone Cottage for which the winery is named was hand-built one stone at a time by the Hellecksons, using the rounded lava stones they found on the property. Just steps from the vines they built a second stone building as their tasting room, which offers stunning views of the North Fork Valley. In the tradition of the Italian "agriturismo", the Hellecksons have converted their original Stone Cottage, which used to be their family home, into a cozy guesthouse.

While the Hellecksons do source some of their grapes from the Grand Valley AVA, they also organically grow 5 acres of Pinot Noir, Merlot, Chardonnay, Pinot Gris, and Gewürztraminer, some more than 35 years old. The vineyards sit at about 6,200 feet (1,890 meters) in elevation and include some of the highest elevation commercial Merlot vineyards in the world. The Stone Cottage wines are expressive and food-friendly. Try the brisk, balanced, and elegantly floral Gewürztraminer, the lemony, mineral-driven Chardonnay, or the supple and smoky Merlot.

The Storm Cellar

Hotchkiss
stormcellarwine.com

The Storm Cellar is the creation of two Denver sommeliers, Steve Steese and Jayme Henderson, who were ready to put their extensive wine knowledge to the test in the high-altitude climes of Colorado's

West Elks AVA. For years, the pair had honed their gardening skills in their home backyard, but the dream of owning a vineyard and starting a winery had taken root in their hearts. Rather than continue to sell the creations of other winemakers, they were keen to begin producing something of their own. In 2017, the two found a property in Hotchkiss that was home to the historic Redstone Vineyard. The parcel, which sits at nearly 6,000 feet (1,829 meters) in elevation, needed a lot of care and hard work. The first couple of years involved an overhaul of the vineyard, which required mended trellising, servicing wind machines to fight frost, transitioning the vineyard from conventional farming practices to organic, and replanting a significant proportion of vines. Since the property had only been a vineyard up until that time, Steese and Henderson also took on the task of converting the estate's old machine shop into a fully-functioning winery.

The original planting from 1987 included 23 acres (9.3 hectares) of own-rooted Old Wente Clone Chardonnay. Over time, all but five acres (2 hectares) of this Chardonnay has been replaced with Riesling, Pinot Gris, and Sauvignon Blanc. But, in cleaning up the vineyard, the couple also discovered a troubling factor: their vineyard was infested with phylloxera. For decades, many have ignored the potential threat of phylloxera in Colorado due to its high altitude and cold temperatures. Because of this, the majority of Colorado's vineyards are own-rooted not only for cost-saving measures but also to mitigate the potential winter damage at the grafting points of the vines. But as discovered in 2016, Colorado's wine regions are not impervious to the root louse (see box, p. 217). The two sommeliers, who had long understood the devastating effects phylloxera imposed on Europe in the late nineteenth century, were now faced with managing it in their vineyards.

As they would learn, the cold weather does help slow the spread of the aphid. Henderson and Steese's approach is simply to replant the vines that have stopped production first and slowly replace the vines that are overcome over time. In 2019 they planted 1,250 grafted Dijon Clone 95 Chardonnay vines and saw positive results over the following year. They plan to add 1,000 new vines every spring, eventually filling all 23 potential acres (9.3 hectares) in the coming years.

In 2017 The Storm Cellar sold its first harvest to existing producers, including The Infinite Monkey Theorem and Carboy Winery, but in 2018, Henderson and Steese released seven offerings for their inaugural vintage. The couple grows Riesling, Chardonnay, Pinot Gris, and

Sauvignon Blanc with the philosophy that white varieties are best suited to the region. There are a few rogue Pinot Noir vines in the existing vineyard, which the couple will likely replant to use for rosé. (They currently make a rosé primarily from Pinot Gris using extended contact on the skins during cold soak.) In general, The Storm Cellar's goal is to have about 80 percent of production come from their estate, with 20 percent coming from other vineyards, including in the Grand Valley.

FRONT RANGE

BookCliff Vineyards
Boulder
bookcliffvineyards.com

BookCliff Vineyards is the creation of John Garlich and Ulla Merz, who first planted their 10-acre (4-hectare) vineyard in Palisade in 1996. In 1999 they officially began releasing wines under the name of BookCliff Vineyards, referencing the towering mesa formation that looms over the Grand Valley. Initially working with Chardonnay, Merlot, and Cabernet Sauvignon, BookCliff has now grown to include an additional 27 acres (10.9 hectares) of vines and is primarily planted to Rhône varieties as well as Graciano and Souzão.

Although the vineyards are located in the Grand Valley, the 5,000-case winery and tasting room are located on the Front Range in Boulder. In 2018, BookCliff hired winemaker Richard McDonald, an Australian native whose experience had not only seen him working in Australia's Riverina wine region, but also in Napa Valley and Sonoma at the likes of The Prisoner Wine Company, Flowers Vineyard and Winery, and Faust. Over the years, BookCliff has cemented a loyal following of Colorado consumers, selling wines from their tasting room, and through retail and restaurant locations. Try the Cabernet Franc Reserve, a consistent performer, with vibrant fruit character and elegant structure; the Tempranillo with notes of dark fruit, tobacco, and leather; or the Ensemble, a soft, fruit-forward Bordeaux blend.

Carboy Winery
Denver
carboywinery.com

Carboy Winery is the brainchild of a few like-minded winemakers and restaurateurs, including Eric Hyatt and Craig Jones, who formed a veritable *négociant* urban winery in Denver in 2014. Carboy's winemakers Tyzok Wharton and Christopher Christensen work with other winemakers and vineyards from prominent wine regions from around the world to source the grapes, juice, and wine to make, blend, and age their wine. In 2016, a second location was opened in the Denver suburb of Littleton and later an extension to the historic Gold Pan Saloon in the ski town of Breckenridge was added. Most Carboy wines are bottled, but wine is also available on tap at each location, leaning on the idea of many craft beer models, which allow customers to take home wine in one-liter refillable growlers. At any given time of year, the different locations feature wines harvested from regions in California, Washington, Oregon, Italy, France, Argentina, and, of course, Colorado. With such a broad model, different wines are regularly rotating through the various locations, making new experiences for repeat customers each time they return.

Creekside Cellars
Evergreen
creeksidecellars.net

Perched on the banks of Bear Creek, this two-decades-old winery and restaurant offers a pastoral setting and respected Colorado-grown wines. Creekside Cellars opened in 2000 under the ownership of William and Anita Donahue. At first, the Donahues produced the wines themselves from their private vineyard in Palisade along with three other Grand Valley growers, but more than a decade ago, they handed the winemaking reins over to Michelle Cleveland.

Originally from Illinois, Cleveland grew up in a farming community, which led her to pursue a degree in agriculture. In 1994, after moving to Colorado, she began taking an interest in the emerging Colorado wine scene. When Creekside opened in Evergreen, she struck up a relationship with the Donahues, and was able to obtain an apprenticeship in 2005. By 2007 she had completed her studies at the University of California, Davis and was offered the position of head winemaker.

Evergreen's location almost midway between Denver and the mountain ski towns makes it accessible to city dwellers. Creekside operates an onsite restaurant, where 99 percent of their wines are sold (the winery produces about 4,000 cases annually). Cleveland considers her wines a collaboration between viticulture and enology. Working closely with vineyard managers throughout the growing season, she aims to make wines that are true to their origin. Though she's partial to a few varieties,

224   THE WINES OF SOUTHWEST U.S.A.

including Syrah, Petit Verdot, and Malbec, Cabernet Franc might be her favorite. Its resilience in the vineyard, coupled with its ability to reflect the high elevation terroir, with bright fruit character and elegant structure, makes it one of the winery's top performers.

The Infinite Monkey Theorem

Denver
theinfinitemonkeytheorem.com

This urban winery got its start in Colorado, helping to put the burgeoning Colorado wine industry on the map. Today, the company is headquartered in Austin, Texas, but continues its Colorado production. See p. 89 for details.

Snowy Peaks Winery

Estes Park
snowypeakswinery.com

Located in the majestic environs of Colorado's celebrated Estes Park, Snowy Peaks Winery was opened in 2005 by Erik and Candice Mohr, two ecologists who added winemaking to their scientific profession. Inspired by the rural scenery of Moraine Park, Snowy Peaks Winery labels all bear an artistic rendering of this frequented valley in Rocky Mountain National Park, which also happens to be where the Mohrs were married.

The Mohrs rely on about a dozen vineyards in Palisade and Paonia as the source for the majority of their grapes, including varieties such as Riesling, Traminette, Viognier, and Tempranillo. But at Burlington Worden Farms, on the eastern slopes, they also grow cold-tolerant varieties including Lacrosse, Frontenac, and Cayuga White. With a selection of more than 30 offerings from dry to sweet, Snowy Peaks has something for everyone. Try Petit Verdot, with fruit-forward notes of plum and vanilla, or the Élevé, a Rhône-style red blend of Syrah, Petite Sirah, and Mourvèdre.

The Winery at Holy Cross Abbey

Cañon City
abbeywinery.com

Far from most other wineries in the state, The Winery at Holy Cross Abbey is located in Cañon City, about 12 miles (19 kilometers) east of the famous Royal Gorge Bridge and Park in Central Colorado. Benedictine monks founded the monastery in 1924, and in 2002 the winery was established, based on the entrepreneurial suggestion of the abbey's Father Paul Montoya. He had visited the vineyards of the Western Slope regions and convinced his fellow monks that if they could make wine, it might be a valuable revenue stream for the abbey's operational expenses.

The homey tasting room is located in an old Arts and Crafts house, built in 1911 as part of the abbey's boys' school. Although the monastery had closed by 2006, the winery remains a thriving business now owned by former New York City accountant Larry Oddo, who relies on winemaker Jeff Stultz, a fourth-generation Coloradan, to produce the winery's dozen or so wine selections.

Holy Cross wines are primarily made from grapes from the Western Slope, as well as those grown at the nearby Cañon City prison complex. The winery helped the prison create a vineyard management program for inmates 18 years ago. Today about 30 percent of the winery's grapes come from the prison's farm, including Merlot, Chardonnay, and Riesling. The winery also sponsors a community blend in which grape-growing residents of the area may contribute grapes to make the Wild Cañon Harvest. This is a blend of a few dozen grape varieties which Stultz shapes into a decent bottling each year.

FOUR CORNERS AND MOUNTAIN

Buckel Family Wine

Gunnison
buckelfamilywine.com

Having earned his stripes as a Colorado winemaker while making wines for Sutcliffe Vineyards for more than a decade, in 2017 Joe Buckel took the next logical step in launching his own brand, Buckel Family Wine, along with this wife, Shamai, and their son and daughter. Buckel's career began in Sonoma, working his way through the labs and vineyards of B.R. Cohn Winery, Cobb Wines, Keith Rutz Cellars, and Flowers Vineyards and Winery. He also partnered with a few friends to create the TAJ Cellars label in Sonoma. But in 2008, he moved to the Four Corners region of Colorado to work for Sutcliffe Vineyards, where he could be closer to his wife's family and spread his wings into the Colorado wine industry.

To start Buckel Family Wine, he relocated his family to Gunnison, a small town on the western slopes of central Colorado, about 30 minutes

226 THE WINES OF SOUTHWEST U.S.A.

south of the famous ski town of Crested Butte. While he also sources fruit from both the Grand Valley and West Elks, he places a lot of confidence in the potential for the Four Corners region of the state, where he can purchase fruit from Canyon of the Ancients Vineyard and Box Bar Vineyard.

Buckel employs a hands-off approach to craft wine using minimal sulfur and no fining and filtration whenever possible. The Buckel Family wines are vibrant yet restrained, with an Old-World elegance that represents what Colorado wine can be in the coming years. Currently, production is at about 2,200 cases annually with the potential to grow.

Try the Cabernet Franc, with notes of juicy blackcurrant, pine forest earth, and spice; the Cinsault, which shines in the Four Corners region, offering bright red fruit and playful minerality; and the soft yet energetic Chardonnay with piercing acidity and Burgundian elegance.

Monkshood Cellars

Minturn
monkshoodcellars.com

One of the more recent additions to the Colorado wine scene, Monkshood is located in Minturn, in the heart of ski country near Vail and Beaver Creek. Minturn, an eclectic old railroad town from the mid-1800s, is replete with quaint, locally-owned restaurants and shops, and, as of 2016, a burgeoning winery. Owner and winemaker Nathan Littlejohn launched the brand following numerous apprenticeships in far-flung regions, including New Zealand, Australia, and Corsica. He later spent three years as a winemaker at Mayacamus Vineyards in Napa Valley. With such rich experiences under his belt, he returned to his native Colorado to develop his own venture, Monkshood, so named for the tall, deep purple wildflower that grows in Colorado's alpine regions. Littlejohn's production is hyper-focused on three things: Syrah, Chenin Blanc, and cider. Revealing his great respect for the natural processes of winemaking, his approach in the cellar is virtually hands-off. Employing native yeast fermentations along with light sulfur additions, and bottling the wines without fining or filtering, Littlejohn's wines are some of the most authentic expressions of Colorado terroir you'll find. You could almost call Littlejohn a wine shepherd rather than a winemaker.

Sutcliffe Vineyards

Cortez
sutcliffewines.com

An outlier in more ways than one, Sutcliffe Vineyards is located on the frontier of Colorado winemaking in the Four Corners region near Cortez. Siloed off along a seemingly endless winding road in the furthest stretches of Colorado's southwest corner, John Sutcliffe, for which the brand is named, is a desperado of sorts in Colorado wine. But he wouldn't have it any other way. Sutcliffe is a true cowboy, and not merely because he dons a broad-brimmed hat and plaid, pearl-snapped shirt, but also because he's spent many years roping cattle and wrestling them to the ground for branding. Indeed, were you to hear the stories of his early ranching days, you'd soon understand there's nothing glamorous or showy about it.

Before his cowboy life, this Welshman had occupations including working his family's racing horse farm in his early years, playing competitive polo as a young man, serving in the British military, and building or managing more than 23 award-winning restaurants, including the famed Tavern on the Green in New York City. But when he moved to Colorado to help work a cattle ranch in the 1980s, his heart was forever captured by the Rocky Mountains.

In 1990, he purchased property in an arid canyon nestled between Sleeping Ute Mountain and the sheer red sandstone walls of the Battlerock. The land, which he aptly named Battlerock Ranch, was riddled with scrub brush and perpetually threatened by mountain lions. He originally envisioned a working cattle ranch, but as construction on his house was nearing completion in 1995, he decided to add a vineyard, as a landscaping aesthetic. Here, the cool, desert mountain air at an elevation of about 5,300 feet (1,615 meters) sees sub-zero temperatures in the winter, offset by hot temperatures well above 100°F (37.7°C) in the summer. The poor, well-drained, sandy-loam soils added to the grape-growing appeal. Sutcliffe soon found that the vines would spark a passion for wine and lead him to establish a winery.

He initially planted Bordeaux varieties but was particularly interested to see how Rhône varieties such as Viognier, Cinsault, and Syrah would fare in the warmer, sandy soils of the Four Corners region. By 1999 he had bottled his first vintage and set about the task of presenting his wines to his trusted confidants in the New York, California, and Colorado restaurant industries. He was encouraged by the response.

228   THE WINES OF SOUTHWEST U.S.A.

He saw his wines finding their way onto numerous high-end wine lists, including the Little Nell in Aspen, Dunton Hot Springs restaurant in nearby Dunton, and Utah's Amangiri resort and spa.

Sutcliffe's wine program has been led by notable winemakers Ben Parsons and Joe Buckel, but has long been supported by assistant winemaker Jesus Castillo, who has also been his most valued vineyard and ranch manager since the inception of Battlerock Ranch. In recent years, Castillo has taken over the reins as head winemaker. Sutcliffe now manages 56 acres (22.7 hectares) of vineyards and also sources fruit from two vineyards in Palisade, producing between 3,000 and 5,000 cases of wine annually.

Try the soft and elegant Chardonnay, which boasts bold citrus notes and subtle oak character; the Cinsault, with warm, spicy notes backed by bright red fruit; or the rich Cabernet Franc with herbaceous undertones.

16

WHERE TO EAT AND STAY IN COLORADO

GRAND VALLEY AVA

Although much of Colorado's tourism remains in the mountains or on the eastern side of the Rockies in the Front Range, recent years have seen an increase in recreational activity to the Western Slope. While the summer harvest of orchard fruit has always been a draw, Colorado wine is also beginning to strengthen the region's appeal. In the Grand Valley, restaurants and accommodations are generally spread out between Grand Junction and Palisade—about 12 miles (19 kilometers) from each other. There's no shortage of fruit stands scattered in between.

Stay

The High Lonesome Ranch
DeBeque
thehighlonesomeranch.com

This mountain retreat is one of the region's more glamorous accommodations. Providing an authentic dude-ranch experience with a Western-luxury feel, the High Lonesome Ranch offers cozy log-cabin accommodations set within a mountain valley in the DeBeque Canyon area. You can also opt for one of the spacious safari-style tents for an authentic "glamping" experience. Outdoor enthusiasts can enjoy horseback riding, kayaking, fly-fishing, and skeet shooting.

Wine Country Inn

Palisade
coloradowinecountryinn.com

Conveniently located in the heart of the Grand Valley regions, just off the highway, the Wine Country Inn is happily situated amid 21 acres (8.5 hectares) of vineyards. Rooms are comfortable and casual, and the lobby hosts complimentary afternoon wine tastings.

Dine

Bin 707 Foodbar

Grand Junction
bin707.com

Fourth-generation Coloradan chef Josh Niernberg has lived in Grand Junction since 2007, running this regional farm-to-table bistro—and promoting Colorado wine along the way. While artfully presented, Niernberg's food is anything but pretentious. Although the menu is perpetually changing, previous offerings such as elk and beet tartare, grilled salmon over rye berry risotto, and rabbit cassoulet give a flavor of what to expect.

Cafe Sol

Grand Junction
cafesolgj.com

With a seasonally changing menu, the fresh American-style fare at Café Sol is locally sourced and rich with soulful flavor. The quiche of the day is always a good option, but the Stevie Ray Baughn salad with kale, winter squash, Brussels sprouts, and blue cheese or the chicken–guacamole panini are worthy selections as well.

The Hot Tomato

Fruita
hottomatopizza.com

Just a short 25 minutes from Palisade in the small town of Fruita, this local staple is worth seeking out. The house-made dough is made daily and proofed for 24 hours before being hand-stretched into a pizza crust. Pizzas are fresh, made-to-order, and delicious. Classic pepperoni never disappoints, but be sure to try their spin on Hawaiian-style pizza with Canadian bacon and slices of fresh summer peaches instead of the standard pineapple.

Palisade Cafe 11.0

Palisade

palisadecafe11.com

This casual cafe in Palisade's downtown area serves up a combination of American, Spanish, and Peruvian cuisine. Tempura-fried cheese curds are an addictively delicious starter and the juicy Pal Burger made with locally-sourced beef makes for a hefty, satisfying meal.

Pêche

Palisade

pecherestaurantcolorado.com

This sleek Palisade locale offers an elevated dining experience using the bounty of the region's produce, meat, and cheese. The seasonal menu offers starters such as pickled plums and beets or charred bone marrow with red onion jam, followed up with main courses that range from the more traditional fried chicken and grits to the exotic kimchi-Peking duck.

Tacoparty

Grand Junction

tacopartygj.com

The sister establishment to Bin 707, this hip locale is all about, you guessed it, tacos. (Actually, you'll find a full menu of Southwestern-inspired cuisine as well.) Tacoparty's format is pretty straightforward. The daily menu offers an ever-changing list of six appetizers, six tacos, and a soft-serve dessert. Simply order your fill and enjoy.

See

Colorado National Monument

Fruita

nps.gov/colm

This park is indeed a national treasure. With more than 300 days of sunshine, the park is a paradise for hikers of all skill levels, offering views of towering red rocks, deep canyons, and large boulders amid Colorado's lush, natural flora. Those who love to cycle can take advantage of Rim Rock Drive, which plays host to the annual Tour of the Moon, a bucket list ride for many cyclists.

Fruit stands

If visiting during the summer harvest season, be sure to take advantage of the numerous fruit stands and shops set up throughout the region.

232   THE WINES OF SOUTHWEST U.S.A.

Many of them offer lunches or picnics to enjoy with a basket of fruit. Among the not-to-miss farm stands are Talbot Farms, Field to Fork, The Produce Peddler, Blaine's Tomatoes, and Okagawa Farms. (See also box below.)

Mount Garfield

Palisade
visitgrandjunction.com

Spanning from Colorado to Utah, the Book Cliffs is the longest continuous cliff face on Earth. Outdoor enthusiasts can hike the Mount Garfield Trail, which covers over 2,000 feet (610 meters) of elevation gain over a four-mile hike (6.5 kilometers). (Be warned: the excursion is rated as "very difficult" due to the steep climb, and lower oxygen at the higher altitude.)

The Colorado fruit basket

Raising orchard fruit in the Rocky Mountains may sound contrary. But as the Colorado River meanders its way from the rugged peaks of the mountains, its banks widen near Palisade. Here, it flows into the verdant Grand Valley, which is framed by flat-top mesas and carpeted by a swath of pear, apple, apricot, cherry, and peach orchards. This region, sheltered by DeBeque Canyon, has a unique microclimate. Today, it is home to the majority of Colorado's vineyards, but it has long been prized as the fruit basket of the state.

The story of Colorado's fruit production dates back nearly 150 years, to the time when irrigation canals were first introduced to the Grand Valley in 1882. Before that, the region had been inhabited by the Ute tribe, who would be forced onto reservations by 1881. Early settlers to the area included George Duke, Samuel Wade, and Enos Hotchkiss, for whom the present-day town in Delta County is named.

Duke, Wade, and Hotchkiss discovered native fruit thriving in the valley at the North Fork of the Gunnison River, indicating the potential for cultivating farmland. Wade returned to his home in Missouri to bring back a few hundred samples of apple, cherry, peach, pear, and apricot trees, along with grapevines and berries to plant in this unexplored region of western Colorado. The following years would yield positive results, prompting other settlers to follow suit. As the system of gravity canals was built throughout the region, the township of Grand Junction was founded by George Crawford at the confluence of the Gunnison and Colorado Rivers, and orchard plantings increased. In 1883,

William E. Pabor founded Fruitvale (now Fruita), following a 12-year study he published, entitled "Colorado as an Agricultural State: Its Farms, Fields, and Garden Lands" (he also penned a book, called *Colorado as an Agricultural State*). The town would become the valley's first fruit tract community. For $500, a farmer could buy 5 acres (2 hectares), 200 fruit trees, and access to water to begin cultivation.

In the following decade, other orchard farm plantings began popping up throughout the region. Those planted closer to the DeBeque Canyon saw greater success, with warmth and protection from the canyon walls. Within a decade of these early plantings, the valley had more than 2,000 acres (809 hectares) of orchard plantings. Pioneering families from the east coast and the American Midwest began packing up and heading west. In 1897, James A. Clark brought his family from Iowa to settle here, establishing Clark Family Orchards. This family-owned operation farms more than 100 acres (40.5 hectares) of orchard fruit in the region today. Like Clark Family Orchards, Talbott Farms is one of the longest-standing farmers in the area. Dating back to 1907, the family pursuit began when peach grower Joe Yager planted his orchard. Years later, his granddaughter would marry a local cattle rancher, Harry A. Talbott, who went on to become a fruit grower.

At the time, mining was the economic driver for much of Colorado, while fruit farming was still a burgeoning industry. But the two would come to thrive in tandem, with Western Slope mining camps providing much of the stimulus for the local fruit market.

By 1909, the region's fruit industry had taken hold. In fact, such was its renown that President William Howard Taft paid a visit for the celebratory Peach Day and spoke of the wonderful fruit of the region. An estimated 15,340 acres (6,208 hectares) had been planted by 1915, and the industry steadily grew over the next century. Although it faced inherent challenges, such as late spring frost, hail, and aggressive winds as well as chlorosis, and pests such as the codling moth, the Colorado fruit industry would continue to grow in the succeeding decades. Its strength would sustain farming communities throughout the Great Depression of the 1930s and be invigorated in the 1940s by the Second World War.

Today, the fruit industry is one of the state's most critical economic pillars, drawing in $41 billion annually. (In the case of peaches alone, compared to California, Georgia, and South Carolina, the Colorado-grown fruit is worth about twice as much per ton.) It is propelled by a few dozen family-owned and operated farms throughout the Grand Valley as well as into Delta County,

234   THE WINES OF SOUTHWEST U.S.A.

which is the West Elks AVA (the industry also thrives in the southwest part of the state). A decent-sized orchard of about 10 acres (4 hectares) can gross upwards of $50,000 annually. The longstanding Talbott Farms in Palisade owns 306 acres (123.8 hectares) of peach orchards, plus another 270 acres-worth of fruit (109.3 hectares) from growers who contract with the farm to sell their peaches. Colorado fruit can be found at grocers nationwide with wild fanfare for Palisade peaches found anywhere from local fruit stands to Whole Foods Markets during the summer months.

Although orchard fruit reigns supreme on the Western Slope, the wine industry's growth has prompted many fruit growers to incorporate vineyards into their plantings. Likewise, grape growers have diversified, growing other fruit alongside their wine grapes, a strategy that has proven to complement revenue, particularly in challenging growing seasons. Fortunately, during a given growing season, the two crops can be managed complementarily, with pruning of peaches followed by vine pruning, shoot thinning, peach harvest, and grape harvest.

As well as selling fruit beyond the region's borders, the orchards of the Western Slope have fostered a dynamic fruit culture—which also includes a rapidly growing wine scene. A booming agritourism business has also taken root, leading to a flurry of new restaurants, bed and breakfasts, and seasonal events that yield a $30 million profit for the local economy. Today, visitors come to the region to take in its recreational beauty, sip from its selection of wines and ciders, and savor its bounty from the myriad farm stands throughout the valley.

Grand Valley fruit stands

Alida's Fruits	Kokopelli Farm Market
Aloha Organic Fruit	Mt. Lincoln Peach Company
Anita's Pantry and Produce	Morton's Organic Orchards
C & R Farms	Palisade Peach Company
Clark Family Orchards	Palisade Produce
Cole Orchard	The Peachfork
Davis Family Farms	Pear Blossom Farms
DeVries Farm Market	Sage Creations Organic Farm
Helmer's Produce	Talbott Farms
High Country Orchards	Taylor Farm and Ranch
Just Peachy	Z's Orchard

WHERE TO EAT AND STAY IN COLORADO 235

North Fork Orchards (Paonia/Hotchkiss)

Antelope Hill Fruit Farm	Excelsior Orchards
Austin Farms	First Fruit Organic Farms
Big B's Delicious Orchards	Happy Belly CSA
Black Bridge Winery	Orchard Valley Farms
Ela Farms	Stahl Orchards

Fruiting seasons

cherries—early June	nectarines—mid-July
early peach varieties—late June	apricots—July
most peach varieties—mid-July	pears—August
end of peach harvest	plums—August
—mid-September	apples—July through October

WEST ELKS AVA

In the West Elks AVA, which is commonly known as North Fork Valley, life slows to a meandering pace, much like the glimmering Gunnison River that flows through the region. With one national and two state parks, plus more than three million acres (1,214,057 hectares) of National Forest nearby, the area is a paradise for cycling, hiking, boating, fishing, and hunting. The small towns of Hotchkiss and Paonia are the primary sites for restaurants and shops, while scattered throughout the region are myriad bed and breakfast and guesthouse options.

Stay

Agape Farm and Retreat

Paonia

agapefarmandretreat.com

Situated a short 6 miles (9.5 kilometers) from Paonia, this relaxing escape is located on a 22-acre (8.9-hectare) property among a pine forest, grassy fields, and verdant gardens, overflowing with cheerful sunflowers and featuring a small Pinot Gris vineyard. Rooms are spacious and comfortable, and there's also a rustic yurt, for those who like the glamping experience. Following a restful slumber, enjoy a delicious home-made breakfast on the outdoor terrace.

Gunnison River Farms

Austin

gunnisonriverfarms.com

Here, visitors can enjoy the comforts of hotel luxury amid a working mountain farm. This recreational oasis has been a working orchard for more than 90 years. Guests can learn about farming orchard fruit while also taking in sweeping views of ponds and rivers backed by red cliff walls, as the farm blends into the dramatic surroundings of deep canyons and open farmland.

Leroux Creek Vineyards and Inn

Hotchkiss

lerouxcreekinn.com

In addition to being among the region's celebrated wine producers, owners Yvon Gros and Joanna Reckert offer a cozy inn on their 54-acre (21.9-hectare) farm. Each room offers spectacular views, and common areas allow for relaxing with a glass of Laroux Creek wine, made from either Chambourcin or Cayuga.

Dine

Flying Fork Café

Paonia

flyingforkcafe.com

Offering a variety of fresh-baked crusty bread, garden salads, house-made pasta dishes, and delicious Neapolitan-style pizzas, this local favorite is the perfect spot to satisfy a craving for Italian fare.

Living Farm Cafe

Paonia

thelivingfarmcafe.com

As you may expect from the name of this homey establishment, farm-fresh, local goods are the primary focus. On a warm day, enjoy a seat on the cozy patio and sample from an array of savory selections, from lamb enchiladas to elk osso buco.

See

Dark Canyon Trail

https://www.fs.usda.gov/recarea/gmug/recarea/?recid=33728

Less strenuous than Mount Lamborn, this scenic trail begins in the

Erickson Springs Recreation Area, about 20 miles (32 kilometers) northeast of Paonia. It follows Anthracite Creek through a deep canyon before switch-backing up the "Devil's Staircase" and crossing the wooded slopes of the remote Ruby Range. Check the Forest Service website for directions to the trailhead.

Mount Lamborn
https://www.fs.usda.gov; search for Mount Lamborn

Paonia's signature hike is the ascent of this 11,396-foot (3,474-meter) peak. The local legend is that the mountain was named for a white, supposedly lamb-shaped scar on its north-facing slope. From the summit, take in views of the West Elks Mountains and the North Fork Valley. Check the Forest Service website for directions to the trailhead.

Paonia's creative district
northforkcreative.org

Paonia is a haven for artists. Spend an afternoon in its downtown arts district, wandering the many turn-of-the-century buildings that house galleries and artists' studios, including Blue Sage Center for the Arts and the Sticks & Stones Gallery. Be sure to take in a movie or live performance at the historic Paradise Theatre.

Southwest literature

When it comes to the history of the American Southwest, in many ways it is hard to know where to draw a line between geological, indigenous, early Spanish, early American, and the era of cowboys, miners, pioneers, railroaders, and outlaws. For a better perspective on the region's heritage, it's best to consult some of the most significant texts written about this storied part of American culture.

Many of these titles offer a glimpse into Mexican and indigenous histories, both of which cultures were overcome with American westward expansion. They reveal the hardship, the heroism, and the perseverance of a region in America that truly has an identity, and is unlike any other place in the country. While a list such as this could indeed be far more extensive, what is included— both non-fiction and fiction—offers a better understanding of the Southwest, which will perhaps help to inform how wine became a part of it.

238 THE WINES OF SOUTHWEST U.S.A.

Overviews of the Southwest

Legends of the American Desert: Sojourns in the Greater Southwest, Alex
  Shoumatoff
The Southwest, David Lavender
Aldo Leopold's Southwest, Aldo Leopold
The Best of Edward Abbey, Edward Abbey
Desert Solitaire, Edward Abbey

Indigenous anthropology

Los Comanches: The Horse People, Stanley Noyes
Life Among the Apaches, John Cremony
Marietta Wetherill: Life with the Navajos in Chaco Canyon, Kathryn Gabriel
Bury My Heart at Wounded Knee: An Indian History of the American West, Dee Brown
An Indigenous Peoples' History of the United States, Roxanne Dunbar-Ortiz

Spanish history in America

Anza's 1779 Comanche Campaign, Ron Kessler
*The Dominquez Escalante Journal: Their Expedition Through Colorado, Utah,
  Arizona, and New Mexico in 1778*, Ted J. Warner
Mexican Frontier (1821–1846): The American Southwest Under Mexico, David J. Weber

Topical history

The Apache Wars, Paul Andrew Hutton
Cochise, Edwin Sweeney
Empire of the Summer Moon, S.C. Gwynne
On the Border with Crook, John G. Bourke
The Trail Drivers of Texas, J. Marvin Hunter
Lamy of Santa Fe, Paul Horgan
The Texas Rangers: A Century of Frontier Defense, Walter Prescott Webb

Fiction

The Border Trilogy: *All the Pretty Horses, Blood Meridian, Cities of the Plain*,
  Cormac McCarthy
Lonesome Dove, Larry McMurtry
The Old Gringo, Carlos Fuentes
Goodbye to a River, John Graves
Apache Gold and Yaqui Silver, J. Frank Dobie
The Best Novels and Stories of Eugene Manlove Rhodes, Eugene Manlove Rhodes

BIBLIOGRAPHY

Adams, Andrew, "Colorado to Survey Phylloxera Risk to Vineyards", *Wines and Vines Analytics*, Sonoma, California, 2017. https://wines-vinesanalytics.com/news/article/184299/Colorado-to-Survey-Phylloxera-Risk-to-Vineyards

The Alcohol and Tobacco Tax and Trade Bureau, "Establishment of Sonoita Viticultural Area", Federal Register vol.49, no.96, May 16, 1984. https://www.ttb.gov/images/pdfs/Sonoita_NPRM.pdf

Barrueta, Christina, *Arizona Wine: A History of Perseverance and Passion*, American Palate: Charleston, South Carolina, 2019

Berg, Eric, "Equal Age for Age: The Growth, Death, and Rebirth of an Arizona Wine Industry, 1700-2000", *Journal of Arizona History*, Tucson, Arizona, 2018

Birchell, Donna Blake, *New Mexico Wine: An Enchanting History*, American Palate: Charleston, South Carolina, 2013

Buchan, Jodi, *Western Colorado Fruit & Wine: A Bountiful History*, American Palate: Charleston, South Carolina, 2015

English, Sarah Jane, *The Wines of Texas: A Guide and a History*, Eakin Press: Fort Worth, Texas, 1995

Holbrook, Christina and Marc Hoberman, *Winelands of Colorado*, Hoberman Publishing: Cape Town, South Africa, 2017

Kamas, Jim, *Growing Grapes in Texas*, Texas A&M University Press: College Station, Texas, 2014

Kane, Russell D., *The Wineslinger Chronicles*, Texas Tech University: Lubbock, Texas, 2012

McLeRoy, Sherrie S., Roy E. Renfro, Jr., Ph.D., *Grape Man of Texas: Thomas Volney Munson and The Origins of American Viticulture*, Eakin Press: Fort Worth, Texas, 2008

Mielke, Eugene A., Dutt, Gordon R., Hughes, Sam K., et al., "Grape and Wine Production in the Four Corners Region", College of Agriculture, University of Arizona: Tucson, Arizona, 1980. https://repository.arizona.edu/bitstream/handle/10150/602124/TB239.pdf?sequence=1

Moloney, Michael, "A History of the Palisade Wine Industry", *Journal of the Western Slope*, Grand Junction, Colorado, 1996

Nickles, Jane, "Welcome to the World, Willcox AVA", winewitandwisdomswe.com, 2016

Parker Wong, Deborah, "Snapshot of New Mexico", *Somm Journal*, 2019. https://www.sommjournal.com/snapshot-of-new-mexico/

Pinney, Thomas, *A History of Wine in America: From the Beginnings to Prohibition*, University of California Press: Berkeley, California, 2005

Pinney, Thomas, *A History of Wine in America: From Prohibition to the Present*, University of California Press: Berkeley, California, 1989

Quinton, Amy, "UC Davis Releases 5 New Wine Grape Varieties", University of California, Davis, 2018. https://www.ucdavis.edu/food/news/uc-davis-releases-five-new-wine-grape-varieties/

Raines, Elaine, "The Beginning of Arizona's Modern Wine Industry", *Arizona Daily Star*, Tucson, Arizona, June 8, 2009

Sexton, Joyce, "History of the Fruit Industry, in Mesa County", Colorado State University College of Agricultural Science, 1992. https://aes-wcrc.agsci.colostate.edu/stations/orchard-mesa/history-of-the-fruit-industry-in-mesa-county

Stephens, A. Ray, *Texas: A Historical Atlas*, University of Oklahoma Press: Norman, Oklahoma, 2010

Sumlin, Callie and Mickelsen, Denise, "The 5280 Guide to Colorado Wine", *5280*, Denver, Colorado, 2019. https://www.5280.com/2019/01/the-5280-guide-to-colorado-wine/

White, Kelli, "The Devastator: Phylloxera Vastatrix and the Remaking of the World of Wine", GuildSomm, Petaluma, California, 2017. https://www.guildsomm.com/public_content/features/articles/b/kelli-white/posts/phylloxera-vastatrix

Wilder, Zibby, "Grapes Expectations: New Mexico's wine industry readies for a historic celebration", *Santa Fe Reporter*, May 21, 2019. https://www.sfreporter.com/news/coverstories/2019/05/22/grapes-expectations

WEBSITES

Arizona Vignerons Alliance: arizonavigneronsalliance.org
Arizona Wine Growers: azwinegrowersassociation.com
Colorado Wine: coloradowine.com
New Mexico climate center: weather.nmsu.edu
New Mexico Wine: nmwine.com
Texas High Plains Winegrowers: txhighplains.com
Texas Wine and Grape Growers Association: txwines.org
Texas Wine Growers: texaswinegrowers.com

ACKNOWLEDGMENTS

That this book was possible was due in no small part to my husband, Myers, and my children, Gus and Ashlyn, who allowed me to drag them around all sorts of winery tasting rooms down the long, seemingly endless roads throughout the Southwest. I realize that what I tried to pass off as adventurous road trips were clearly not conventional family vacations, but hopefully, you had a whole lot of fun along the way. I'm forever grateful for your willingness to let me "get out there and see", as well as your patience with me as I clacked away at my computer keyboard for hours on end. You are truly my team.

To God, in whom I find my strength, to my mom, for always helping to fill in the gap while I'm knee-deep in a deadline, and to my dad, for miraculously instilling within me an unending love for history, the outdoors, and the culture of the amazing region in which we live, thank you.

A very special thank you to Denise Clarke, who encouraged me to start covering Texas wine more than a decade ago, and to Pat Sharpe, for agreeing to let me do so with *Texas Monthly* magazine. This journey has been enormously fulfilling.

I owe so much to the Texas wine industry, which has given me boundless content to explore and write about for the past decade. I look forward to what the next decade will bring. In addition, I'm thankful for the warm welcome I received from all in the Arizona, New Mexico, and Colorado wine industries while researching this book. I am so grateful for the generous time, hospitality, and extensive information given to me by the High Plains Winegrowers, the Arizona Vignerons Alliance, the New Mexico Wine and Grape Growers Association, the Colorado

244 THE WINES OF SOUTHWEST U.S.A.

Wine Industry Development Board, Andy and Lauren Timmons, Nick and Katy Jane Seaton, Fritz Westover, Dr. Ed Hellman, Jim Kamus, Ashley Hausman MW, Jason Hisaw, Dr. Gill Giese, Doug Frost MS MW, Todd and Kelly Bostock, Maynard James Keenan, Pavle and Carla Milic, Kent and Lisa Callaghan, and Kaylah Kolasinski.

To my wine "study bunnies", including Denise Clarke, Matt McGinnis, Margaret Shugart, Steve Alley, Kristi Willis, Jaime Chozet, Mandi Nelson, and Amé Brewster: when it comes to all things wine, you continue to help keep me sharp and engaged.

I'm grateful to a few non-wine contributors to this project, including T.J. Tucker for his handy design skills, John Woolmington, for his inspiring knowledge of the history of the American Southwest, Kenny Braun for his fabulous photographic eye, and Kate Rodemann, Stacy Hollister, Tavaner Sullivan, and Shana Clarke for keeping me on track and encouraging me along the way.

I want to thank the Infinite Ideas team of editors and staff, including Richard Burton, Rebecca Clare, and Sarah Jane Evans MW. And finally, my sincerest gratitude to James Tidwell MS, who offered me the opportunity to take on this project. Your confidence in my ability to do it is both humbling and encouraging. Thank you.

INDEX

1725 Vineyard 41
1851 Vineyards 97–8

Ab Astris 81–2
Abbott Vineyard 41
acid mine drainage 199–200
Adams, Matt 107
Adams, Penny 102
Agape Farm and Retreat, Paonia 235
Aglianico 7
   Arizona 150, 159, 171, 172, 173, 176, 180
   New Mexico 26, 29, 34, 35
   Texas 85, 86, 89, 96, 97
Agostino Block 171
Alamo 125
Alamosa Wine Cellars 76
Albariño 76, 77, 96, 117, 159, 171
Al Buhl Memorial Vineyard 167, 176, 178
   Keenan/Caduceus Cellars 137, 148, 159,
      170, 172
Albuquerque 44–6
Alcantara Vineyards and Winery 175, 176
Alfred Eames Cellars 218–19
Alicante Bouschet 74, 77, 93, 100, 118
Alta Loma Vineyard 106
Altgelt, Ernst Hermann 63
Amaro Winery 19, 27
Apache 132–4
Apache Pass, Battle of 134, 169
Apache Wars 134
Aramella 211
Arinto 93
Arizona 1–2, 4, 141–2

challenges 79, 152–3
climate 142–3
Frost on 8
Grand Canyon 144–5
grapes 150–2
history 131–9
map 146
mavericks and pioneers 149–50
producers
   Phoenix area 180–1
   Sonoita 155–63
   Verde Valley 169–80
   Willcox 164–9
regions 145
   Sonoita 145–7
   Verde Valley 148–9
   Willcox 147–8
travel
   Scottsdale 183–7
   Sonoita 187–90
   Verde Valley 191–3
   Willcox 190–1
Arizona Farm Winery Act (1982) 136
Arizona Stronghold Vineyards 138, 153, 159,
   165, 169–70, 179
Arizona Vignerons Alliance 163–4, 171, 172
Arizona Vignerons Alliance Symposium 187
Arizona Wine Growers Association 138, 163
Arneis 34
Arrandell 41
Arteaga, Antonio de 12
Asmundson, Phil and Kim 157–8
Asylum, The, Jerome 193

245

246   THE WINES OF SOUTHWEST U.S.A.

Auler, Ed 59, 86–7, 88, 100
Auler, Susan 59, 86–7
Austin Food & Wine Festival 87

Baco Noir 37, 41
Badman, Charleen 184, 185, 186, 187
Bagley, Daniel and Samuel 132
Baird, David 178
Balcones Escarpment 70
Baldwin-Eaton, Jenne 216
Barbera
    Arizona 136, 156, 159, 172, 178, 180
    Colorado 216
    New Mexico 29, 34, 35
    Texas 82
Barroul, Louis 168
Batek, Mike and Denise 89
Battle of Apache Pass 134, 169
Battle of San Jacinto 125
Bear Vineyards/Tio Pancho Vineyard 93, 108
Beat Generation 47
Bechard, Joe 175
Becker, Bunny 82
Becker, Richard 82, 111
Becker Vineyards 62, 82, 103
Becknell, William 15
Bell Mountain 75, 95
Bell Mountain Winery 95
Bending Branch Winery 83
Berg, Erik 131, 132, 135
Berman, Aaron and Meredith 89–90
Bespoke Inn, Scottsdale 183
Bianco, Chris 184
Big Tex BBQ Restaurant, Willcox 191
Bin 707 Foodbar, Grand Junction 230
Bingham, Cliff and Betty 106
Bingham Family Vineyards 76, 85, 106
Black Hamburg 198
Blackmon, William ("Bill") 102–3, 104
Black Muscat 32
Black Spanish 74–5
    Texas 3, 57, 74, 75, 77
        east 117, 119
        Hill Country 91, 92
        south 114, 115
        west 112, 113
Blackwater Draw Vineyards 107
Blanc Du Bois 78
    Texas 77
        east 117, 119

Hill Country 91, 92
    south 114
    west 113
Bloomoon Vineyard 169
Blue French *see* Black Spanish
Bodega Pierce 148, 149, 164–6, 178
Bonarrigo, Paul Mitchell and Karen 116
Bonarrigo, Paul V. and Merrill 59, 114–16
Bonita Springs Vineyard 170, 179
BookCliff Vineyards 215, 222
Bosque Farms 37
Bostock, Frank 160
Bostock, Kelly 160, 164, 181
Bostock, Paula 160
Bostock, Todd
    Arizona Vignerons Alliance 164
    Dos Cabezas WineWorks 137, 138, 158–60,
        164, 181
    Frost on 8
    Garage-East 181
    as mentor 161
Bowden, Tony 112
Box Bar Vineyard 226
Brandecker, Ed and Roberta 112
Branded H Vineyards 96
B.R. Cohn Winery 225
Brennan, Pat and Trellise 110–11
Brennan Vineyards 110–11
Brilliant 117
Brinkmeyer, Chris 175
Bronk, Joseph 36
Brundrett, Chris 102, 103, 104
Buckel, Joe 225–6, 228
Buckel, Shamai 225
Buckelew, Brad 94
Buckel Family Wine 225–6
Buena Suerte Vineyard 137, 155
Buhl, Al 137, 150, 158, 159, 167, 169
Bureau of Alcohol, Tobacco, and Firearms 87
Burklee Hill 104
Burlington Worden Farms 224
Burning Tree Cellars 170, 175, 179
Burran, Ronny and Gale 106

Cabernet Franc
    Arizona 150, 151, 155, 157, 159, 161
    Colorado 210
        Four Corners 226, 228
        Front Range 222, 224
        Grand Valley 208, 215, 216

INDEX 247

New Mexico 18, 26, 28, 41
Texas 77, 82, 84, 116, 118
Cabernet Grill, Fredericksburg 122
Cabernet Sauvignon
Arizona 136, 150
Sonoita 155, 157, 163
Verde Valley 170, 171, 172, 178
Colorado 208, 210, 214, 215, 216, 222
New Mexico 18–19, 28, 38, 39
Mesilla Valley 27, 32
Middle Rio Grande Valley 37
Mimbres Valley 26, 33
phylloxera 61
Texas 76, 77
east 117, 118
High Plains 103, 106
Hill Country 84, 86, 95, 97, 100, 101
south 114
west 111
Caduceus Cellars 149, 153, 170–3, 175
Al Buhl Memorial Vineyard 137, 148, 172
Arizona Vignerons Alliance 164
Luna Rossa Winery 34
Southwest Wine Center 172, 175
Cafe, The, Sonoita 188
Cafe Sol, Grand Junction 230
Calais, Benjamin 84, 88
Calais Winery 84, 88
caliche soils 67
Callaghan, Harold 155–6
Callaghan, Karen 155–6
Callaghan, Kent 8, 137, 155–7, 158, 161, 164
Callaghan, Lisa 156, 164
Callaghan Vineyards 137, 147, 151, 153,
155–7, 164, 168
Callahan, James 158, 161–2, 167
Camp Comfort, Comfort 121
Camp Lucy, Dripping Springs 121
Canada, Dwayne and Brenda 107
Canada Family Vineyards 82, 107
Cañon City prison 225
Canyon of the Ancients Vineyard 226
Caprock Escarpment 69–70
CapRock Winery 62, 104, 109
Carboy Winery 221, 222–3
Cárdenas, García López de 144
Carignan 82, 88, 90, 109, 110, 180
Carl, Prince of Solms-Braunfels 63
Carleton, James 134
Carlson, Parker and Mary 213

Carlson Creek Vineyards 176
Carlson Vineyards 213–14
Casa Rondeña 27
Castillo, Jesus 228
Cayuga 219, 224, 236
Cellar Uncorked, The 39
Centanni, Jason 105–6
Chambourcin
Colorado 3, 210, 211, 219, 220, 236
New Mexico 3, 37
Texas 117, 118
Champanel 117
Charbono 83
Chardonnay
Arizona 150, 165, 167, 170
Colorado
Four Corners 226, 228
Front Range 222, 225
Grand Valley 213, 216
West Elks 220, 221
New Mexico 28, 39, 41
Mesilla Valley 32
Middle Rio Grande Valley 35, 36, 37
Mimbres Valley 26, 33, 34
Northern New Mexico 28
Texas 8, 86, 88, 95, 100, 105
Chasselas 61
Chateau Tumbleweed 149, 153, 175–7, 178
Chavez, Ben 17
Chenin Blanc
Colorado 226
New Mexico 33, 34 39
Texas
east 118
High Plains 102, 106, 110
Hill Country 86, 87, 88, 93
Cherokee Creek Vineyard 102
Chiricahua Apache 132–4
Chiricahua Ranch Vineyard 169
chlorosis 113
Christensen, Christopher 223
Cimarron Vineyard 137–8, 159, 169, 176,
178, 181
Cinsault
Arizona 180
Colorado 210, 226, 227, 228
New Mexico 37
Texas 76, 77, 88, 90, 109, 110
Civil War 56, 63, 134
Clairette Blanche 81, 150

248 THE WINES OF SOUTHWEST U.S.A.

Clark, James A. 233
Clark, William 197
Clark Family Orchards 233
Cleveland, Michelle 223–4
climate change 4
Cobb, Jennifer, Chris, Diane, and Reed 90
Cobb Wines 225
Cochise, Chief 133–4, 169
Cocina de la Sirena, Lubbock 127
Cold Creek Vineyards 6
Colibri Vineyards 162, 170, 176, 180
Coloradas, Mangas 134
Colorado 1–3, 4, 203
    challenges 79, 211–12
    climate 204–5
    Department of Public Health and
        Environment 200
    Frost on 8
    fruit 231–5
    Geological Survey 199
    geology 205–6
    grapes 210–11
    history 197–201, 232–3
    map 207
    mining 233
    phylloxera 217
    producers
        Four Corners and Mountain 225–8
        Front Range 222–5
        Grand Valley 213–18
        West Elks 218–22
    regions 206, 209
        Four Corners 209
        Front Range 209–10
        Grand Valley 206–8
        West Elks 208–9
    travel
        Grand Valley 229–35
        West Elks 235–7
Colorado Cellars 214
Colorado Mountain Vineyards 201, 214
Colorado National Monument 231
Colorado State University 200
Colorado Wine Industry Development Board
    201
Colterris 214–15
Comanche Vineyard 111
Contador, Guillermo 32
Continental Divide 205, 206
Convent 58–9

Cooper, Clarence 31
Coronado, Francisco Vázquez de 144
Corrado, Luca 172
Cortés, Hernán 12
Cotton Gin Village, Fredericksburg 122
Cottonwood Hotel 192
Coturri, Tony 85, 91
Counoise 88, 94, 158, 159, 180
Coyote Cafe, Santa Fe 49
Crafted at Hotel Chaco, Albuquerque 45
Crawford, George A. 198, 232
Creekside Cellars 223–4
Crema Craft Kitchen and Bar, Cottonwood
    193
Crimson Cabernet 83
Criolla see Mission
Crowell, Todd 97, 101
Crowson, Henry and Kenny 85
Crowson Winery & Tasting Room 71, 85
Cuadra, Sergio 87–8
Curtis, John and Susan 1068
Cynthiana 117

Dai Due, Austin 122
Dancing Apache Ranch 138
D'Andrea, Marco 34
D'Andrea, Paolo 19, 26, 29, 32, 33–4, 137
D'Andrea, Sylvia 19, 34, 35
D'Andrea family 31, 36
D.A. Ranch Vineyard 176
Dark Canyon Trail 236–7
David Bruce Winery 179
Davis, William 15
De Chaunac 38
Deep Sky Vineyards 147, 157–8, 170, 176,
    178
Deep Sky Winery 162, 175
Delicatessen 117
Dell Valley Vineyards 105
de Wet, Pierre 118–19
D.H. Lescombes Family Vineyards 29, 32–3
Diamanté Doble Vineyard 96, 107
DiChristofano, Frank 137, 158, 159
Dickson, Lewis 85, 91–2
Dolcetto
    New Mexico 26, 29, 32, 34, 35
    Texas 96, 102, 119
Domestic Farm Winery Bill 138
Donahue, William and Anita 223
Dos Cabezas Vineyard 169, 170

INDEX 249

Dos Cabezas WineWorks 137, 147, 153, 158–60, 187
  Arizona Vignerons Alliance 164
  Bostock 137, 138, 158–60, 164, 181
  Next Door @ Dos Cabezas 188
  Pillsbury 167
  Pizza Truck, Sonoita 189
  Retreat Bed & Breakfast, Willcox 190
Dos Padres Vineyards 176, 180
Downtown Clifton Hotel, The, Tucson 188
downy mildew 78
"Drink Local" movement 7
DuBois, Emile 78
Duchman, Stan and Lisa 85
Duchman Family Winery 85–6
Duckhorn Wine Company 215
Dudley, Kara 100
Duke, George 232
Dunham, Curt 166
Dunkelfelder 74
Dunn, J.H. 59
Durif 180
Durrett, Marnelle de Wet 119
Dust Bowl 47
Dutt, Gordon 135–6, 137, 162–3

Echo Canyon Vineyard & Winery 138, 165, 179
Edge of the Vine Vineyards 176
Edwards Plateau 70
Eisenhower, Dwight D. 47
Eliphante Block 171–2
El Paso (grape) *see* Black Spanish
El Rey Court, Santa Fe 48
Emerald Riesling 86, 136
Enchanted Rock, Texas 70
Enchanted Rock State Natural Area 125
Enchanted Rock Vineyard 93, 108
Engle Vineyard 39
English, Tommy and Jana 104
English Newsom Cellars 104–5
Enoch's Stomp 117
Erath, Dick 137–8, 159
Erath Winery 137–8
Escondido Valley 75
Evie Mae's Pit Barbecue, Wolfforth 127
Exposition, 1893 (Chicago, Illinois), Exhibit of American Grapes 61

Fairhaven 117–18

Fall Creek Vineyards 59, 86–8, 100, 115
Farm at South Mountain, The, Phoenix 184
Farmhouse Vineyards 82, 98, 107
Farrell, Tiffany 114
Farside Bistro, Cottonwood 193
Favorite 117
Feast, Tucson 189
Federal-Aid Highway Act 47
Felton, Bo 215
Fiandaca, Peggy 166
fiction 238
Fire Vineyard 41
Five Lions Vineyard 92
Flame Tokay 198
Flat Creek Estate 62
Flatiron, The, Jerome 193
Flowers Vineyards and Winery 225
Floyd, Ronnie and Bobby Jo 106
Flying Fork Café, Paonia 236
FnB Restaurant, Scottsdale 184, 185–7
Forester, C.O., Jr. 58
Four Corners 209
  grapes 210
  producers 225–8
Four Corners Project 214
Four Corners Regional Commission 135, 136, 200
Four Eight Wineworks 165, 171, 172, 176, 177–8
Four Seasons Rancho Encantado, Santa Fe 48, 50
France
  Colorado history 197
  New Mexico history 15, 18–19
  phylloxera 59–62
  Texas history 55, 59
Fredericksburg in the Texas Hill Country 75
Fredericksburg Winery 62
French Colombard 136
French Connection Wines 88
Friedrich, Prince of Prussia 63
Fritsche, Josh 98
Front Range 209–10
  producers 222–5
Frontenac 211, 224
Frost, Doug 6–8
fruit 231–5
Furgeson, Anthony and Traci 107

Gago, Peter 172

250 THE WINES OF SOUTHWEST U.S.A.

galleries, Santa Fe 50
Gamay 161
Garage-East 180–1
Garduños, Albuquerque 45
Garibaldi, Giuseppe 15
Garlich, John 222
Gatlin, Dan 76
Germany
  New Mexico history 18, 19
  Texas history 63
Gewürztraminer
  Colorado 209, 210, 213, 216, 219, 220
  New Mexico 28
Ghost City Inn, The, Jerome 192
Gimino, Victoria 88
global warming 4
Glomski, Eric 159, 162, 167, 169–70,
  179–80
Glomski, Terry 170, 179
Gold Rush 15, 199
Graciano
  Arizona 150
    Sonoita 156, 157, 159, 160
    Verde Valley 176
    Willcox 165, 166
  Colorado 222
  Texas 96, 101
Grand Canyon 144–5
Grand Valley 206–8
  challenges 211
  grapes 210
  producers 213–19
  travel 229–35
Grape Creek Vineyards 62
Grayson College, Texas 61
Great Divide 205, 206
Grenache
  Arizona 137, 150
    Sonoita 156, 157, 160, 161, 162
    Verde Valley 172, 173, 179, 180
    Willcox 166, 167, 169
  Colorado 208, 210
  New Mexico 29
  Texas 88, 90, 110
Griffiths, Jesse 122
Gros, Yvon 219, 236
Gruet 8, 19, 27, 35–7, 39, 48
  Bubble Bar at Level 5, Albuquerque 46
Gruet, Gilbert 19, 32, 35, 36
Gruet, Laurent 19, 33, 36

Gruet, Nathalie 19
Gruet family 31
Grüner Veltliner 41
Gunnison River Farms, Austin 236

Haak, Gladys 114
Haak, Raymond 75, 114
Haak Vineyards & Winery 75, 78, 114
Hamilton, Scott, Sherrie, and Kyle 216
Hammelman, Rob and Sarah 164, 168–9
Hanson, Lance and Anna 219
Headstream, Tom 112
Heartwood Cellars 175, 178
Helleckson, Brent and Karen 220
Hellman, Ed 112
Henderson, Jayme 220–2
Hendricks, Jeff 175–6
Herbemont 113
Herbemont, Nicholas 74
Hester, Randy 98
Hewitson 168
Hidalgo, Treaty of (1848) 132
High, Scott and Theresa 214–15
High Country Orchards 215
High Lonesome Ranch, The, DeBeque 231
High's Cafe and Store, Comfort 123
Hill, Chace and Elizabeth 104, 107
Hobbs, Paul 87
Hog Heaven Vineyards 86
Home Vineyard 179–80
Hoover Valley Vineyard 81
HossFly Vineyards 85
Hotchkiss, Enos 232
Hotel Albuquerque at Old Town 45
Hotel Chaco, Albuquerque 45, 46
Hotel Encanto, Las Cruces 43
Hotel St. Francis, Santa Fe 48
Hotel Valley Ho, Scottsdale 183–4
Hot Tomato, The, Fruita 230
House Mountain Vineyard 176, 179–80
Hunter Vineyards 103
Hyatt, Eric 222–3
Hye Meadow Winery 89

Ibarra, Francisco de 11
Infinite Monkey Theorem, The (IMT) 221
  Austin 89–90
  Denver 224
Institut National Agronomique, Montpellier 60
Inwood Estates 76

INDEX 251

Italy
  New Mexico history 15, 18, 19
  Texas history 57, 112
Ivancie, Gerald 200
Ivancie Cellars 200
Ives, Joseph Christmas 144

Jack (grape) *see* Black Spanish
Jack Rabbit Hill Farm 219
Jacquez *see* Black Spanish
Jannusch, Justin 215
Jaqué *see* Black Spanish
Jaramillo Vineyards 27
Javelina Leap Vineyard & Winery 138, 178
Jefferson Cup Invitational Wine Competition
  6, 7
Jennings, Waylon 126
Johnson, Jim 76
Johnson, Karen 76
Johnson, Michael 38
Johnson, Nathan 38
Johnson, Patrick 38
Jones, Craig 222–3
Judith's Block 171

Kasha-Katuwe Tent Rocks National
  Monument 50
Keeling-Schaeffer Vineyards 148
Keenan, Jennifer 164
Keenan, Maynard James 138, 159, 170–3, 175
  Al Buhl Memorial Vineyard 137, 148, 159,
    170, 172
  Arizona Stronghold Vineyards 169–70, 179
  Arizona Vignerons Alliance 164
  Four Eight Wineworks 177
  Frost on 8
  Southwest Wine Center 172, 174
Keith Rutz Cellars 225
Kerouac, Jack, *On the Road* 47
Kessler, Guenhael 112
Kiepersol Estates Winery 118–19
Kino, Eusebio Francisco 131
Klein Vineyards 93
Koegelenberg, Altus 117
Koistinen, Kim 176
Kosta Browne Winery 161
Kotze, Lood 110
Kral, Jon 117
Krick Hill Vineyards 107
Kubacak Vineyards 96

Kuhlken, David 92, 95
Kuhlken, Jeanine 94–5
Kuhlken, Julie 95
Kuhlken, Larry 94–5, 96
Kuhlken Vineyards 95
Kuhlman Cellars 90

La Chiripada Vineyards 28
La Chiripada Winery 18, 38
La Crescent 3, 211
Lacrosse 224
La Cruz de Comal 85, 91–2
La Diosa Cellars, Lubbock 127–8
La Fora, Nicolas de 14
La Hacienda de Sonoita 188
Lahey Vineyards 96, 107
Lamy, Jean-Baptiste 15
La Pradera Vineyards 93, 94, 107
La Salle, René-Robert Cavelier, Sieur de 55
Las Cruces 43–4
La Unica Restaurant & Taqueria, Willcox 191
La Viña Winery 18, 27, 31–2
LDV Winery 166
Leahy, Jonathan 82
Lee, Adam 119
Lemberger 213, 214
Lenoir *see* Black Spanish
Leon Millot 37, 38, 41
Lepard, Russell 107
Lepard Vineyards 107
Leroux Creek Vineyards and Inn 219–20, 236
Lescombes, D.H. 33
Lescombes, Emmanuel 19, 33
Lescombes, Florent 19, 33
Lescombes, Hervé 18, 29, 32–3
Lescombes family 26, 31
Lescombes Winery & Bistro, Las Cruces 27,
  33, 44
Level 5, Albuquerque 46
Levy, Mitch 170, 175, 179
Lewis, Doug 92–3
Lewis, Meriwether 197
Lewis Wines 71, 92–3
Lignan Blanc 171
Listan Blanco 16
Listan Negro/Prieto *see* Mission
literature 237–8
Littlejohn, Nathan 226
Living Farm Cafe, Paonia 236
Llano Estacado 69

252 THE WINES OF SOUTHWEST U.S.A.

Llano Estacado Winery 34, 59, 105–6, 115
Llano Uplift 70
Loew, Bob 158
Lomanto 117, 118
Lordsburg Vineyard 33
Los Milics Winery 147, 160–1, 187
Los Pinos Ranch Vineyards 119
Lost Draw Cellars 93–4, 103
Lost Draw Vineyards 93, 94, 107
Louisiana Purchase (1803) 197
Luckenbach 126
Lulienga 171
Luminaria, Santa Fe 49
Luna Rossa Pizzeria, Las Cruces 35, 44
Luna Rossa Winery 27, 33–5, 41
  D'Andrea 19, 26, 29, 33–4, 36
Lurton, Jacques 87

Magowan, Don and Katherine 158
Maier, Benjamin 19
Maier, Bernd 19, 32, 33
Mainzer Adelsverein 63
Majek Vineyard & Winery 98
Malaga 198
Malbec
  Arizona 150, 157, 158
  Colorado 208, 224
  Texas 77, 90, 97, 111, 114, 118
Malvasia Bianca
  Arizona 136, 137, 150, 151
    Sonoita 156, 157, 159
    Verde Valley 171, 172, 180
    Willcox 165, 166, 167, 168
  New Mexico 34
  Texas 85
Manigold, Ed and Madeleine 100
maps
  Arizona 146
  Colorado 207
  New Mexico 25
  Southwest U.S.A. 5
  Texas 66
Marcus, Jon 138, 158, 165
Marechal Foch 41
Marquette 3, 211
Marsanne
  Arizona 150, 156, 161
  Texas 77, 88, 104, 106
Marselan 179
Marshall, James W. 199

Martin, Francisco 41
Martin, Laura 96
Martin, Seth 96
Martini, Louis 87
Marzo Block 171, 172
Matyskiewski, Patric 218
Mayacamus Vineyards 226
McCrary, James Madison and Ella 111
McDonald, Richard 222
McEachern, George Ray 59
McGlaughlin, Dan and Jeanie 108
McHenry, Mike 101–2
McIntyre, Dave 7
McNabb, Duncan 92–3
McPherson, Clinton ("Doc") 59, 76, 86, 105,
  108, 109
McPherson, Kim 59
  CapRock Winery 104, 109
  Frost on 8
  Infinite Monkey Theorem 90
  McPherson Cellars 109–10
  as mentor 109, 111
McPherson, Sylvia 127
McPherson Cellars 108–10
Meador, Regan and Carey 99
Meow Wolf, Santa Fe 50
Merkin Vineyards 148, 158, 170–3, 175
  Tasting Room & Osteria, Cottonwood 149,
    193
Merlot
  Arizona 136, 150, 155, 163, 170, 178
  Colorado 210
    Front Range 222, 225
    Grand Valley 208, 213, 214, 215, 216
    West Elks 220
  New Mexico 19, 28, 32, 33, 41
  Texas 77
    east 118
    High Plains 103
    Hill Country 82, 84, 86, 90, 95, 101, 104
Merz, Ulla 222
Mescal Canyon Retreat, Clarkdale 192
Mesilla 44
Mesilla Valley 24, 26–7, 75
  producers 31–2
Messier, Tom 175
Messina Hof Wine Cellars 75, 114–16
Meunier 41
Meusebach, Baron Ottfried Hans von (John
  O.) 63

Mexican–American War 63, 125, 132, 133
Mexico
  Arizona history 133
  literature 237
  New Mexico history 12, 15, 16
  Texas history 56, 112–13, 125
Middle Rio Grande Valley 24, 27
  producers 35–7
Mielke, Eugene 135
Milic, Carla 160
Milic, Pavle 160–1, 167, 184, 185–7
Milne, Kim 87
Mimbres Valley 24, 25–6, 28
  producers 32–5
mining history, Colorado 199–200
Mission (grape) 14, 16–17, 163
Missionara see Mission
missions
  Arizona 131, 133
  New Mexico 11, 12, 14–15, 16
  Texas 55–6, 124–5
Mitchell, Roy 59
Moench, Angela Downer 101
Mohr, Erik and Candice 224
Monchâtre, Christian 46
Mondavi, Robert 87
Monkshood Cellars 226
Montag, Noelle 108
Montepulciano
  Arizona 138, 161, 176, 179
  New Mexico 29, 34, 35
  Texas 81, 82, 85, 89, 96, 104
Montoya, Paul 225
Mortensen, John 78
Mount Garfield 232
Mount Lamborn 237
Mourvèdre
  Arizona 137, 150
    Sonoita 156, 158, 160
    Verde Valley 170, 171, 173, 178, 180
    Willcox 167
  Colorado 208, 210, 224
  New Mexico 18, 26, 29, 33
  Texas 76, 77
    east 118
    High Plains 106, 109, 110
    Hill Country 82, 88, 90, 93, 95, 96, 104
    west 111
Mueller, David 104–5
Munson, Thomas Volney 57–8, 60–1, 74, 76, 117

Munson, Tom 112
Muscadine 57
Muscanal 58
Muscat 78, 159, 198
Muscat of Alexandria 111, 118
Mustang 57, 113

Narra, Nikhila, Koteswari, and Nagarjun 107
Narra Vineyards 81, 85, 107
Nation, Briana 178
National Grape Registry 74
National Monuments
  Colorado 231
  Grand Canyon 144–5
  Kasha-Katuwe Tent Rocks 50
  White Sands 23
National Parks
  Colorado 203
  Grand Canyon 145
  White Sands 23
Native Americans
  Arizona 132–4, 144, 169
  Colorado 197, 198, 232
  literature 237, 238
  New Mexico 12, 14, 36, 46
  Texas 55–6, 63, 70
  turquoise mining 51
Nebbiolo 35, 168, 171, 172
Negra Criolla see Mission
Negroamaro 34, 35, 89, 172, 174
Nelson, Michael 81
Nelson, Willie 126
Nelson family 108
Nero D'Avola 34, 111
New Mexico 1–3, 4
  challenges 29–30, 79
  climate 21–2
  Frost on 8
  geology 22–4
  grapes 28–9
  history 11–20, 56–7, 124
  map 25
  producers 31, 38–41
    Mesilla Valley 31–2
    Middle Rio Grande Valley 35–7
    Mimbres Valley 32–5
  regions 24
    Mesilla Valley 24, 26–7
    Middle Rio Grande Valley 24, 27
    Mimbres Valley 24, 25–6, 28

254 THE WINES OF SOUTHWEST U.S.A.

Northern New Mexico 27–8
travel
  Albuquerque 44–6
  Las Cruces 43–4
  Santa Fe 46–51
New Mexico State University
  Fabian Garcia Research Center 20
  School of Agriculture 20
New Mexico Vineyards 33, 34
New Mexico Wine 29, 30
New Mexico Wine Growers Association 20
Newburg Vineyard 111
Newsom, Janice 107
Newsom, Neal 76, 107
Newsom, Steve 105
Newsom Vineyards 81, 84, 93, 107
Next Door @ Dos Cabezas, Sonoita 188
Niernberg, Josh 230
Nitodal 117
Nobility Society of Mainz 63
Noble, Jesse 172
Nogalero Estate Vineyard 96
Noisy Water Winery 31, 38–9
Northern New Mexico 27–8
Norton 101, 117, 210
Nueces Massacre (1862) 63

Oak Creek Vineyard and Winery 138
Oberhelman, Bob 95
Ochota, Taras 172
Oddity Wine Collective, The 178
Oddo, Larry 225
Offill, Tony 103
Ohio (grape) *see* Black Spanish
O'Keeffe, Georgia 50
Old Garden Mourvèdre, Adelaide 168
Olea, Fernando 49
Oñate, Don Juan de 11, 12, 26, 75
One Elm Vineyard 107
One Way Vineyard 96
Orchard Mesa Wine Company 214
Osso, Gio 183
Oswald Vineyards 85
Ottmers, Troy 94
Otto's, Fredericksburg 123
Overton Hotel, The, Lubbock 127

Pabor, William E. 233
Padberg, Jesse, Chris, Michele, and Liliana 40–1
Paddack, Michael and Barbara 107

Page Springs Cellars 138, 149, 167, 170, 175–6, 179–80
País *see* Mission
Palisade Cafe 11.0 231
Palomino Blanco 16
Pantry Restaurant, Santa Fe 48
Paonia's creative district 237
Parlour, The, Johnson City 123
Parr, Robb and Dilek 108
Parr Vineyards 93, 108
Parsons, Ben 228
Pass Wine 57
Patagonia Lake State Park 190
Patagonia-Sonoita Creek Preserve 190
Pattillo, Sheri and Pat 88
Pecan Street Brewery, Johnson City 123
Pêche, Palisade 231
Pecorino 35
Pedernales Cellars 92, 94–6
Penna, Mark 85
Peralta, Don Pedro de 46
Perez, Amulfo 119
Perissos Vineyard and Winery 96–7
Perry, Ron 59, 115
Petersen, Alfred Eames and Devin Eames 218–19
Petite Sirah
  Arizona 136, 150
    Sonoita 158, 159
    Verde Valley 178, 180
    Willcox 165, 166, 167
  Colorado 224
  Texas 77, 81, 83, 90, 96
Petit Manseng 150, 161
Petit Verdot
  Arizona 150, 156, 159, 161, 168
  Colorado 224
  New Mexico 26, 41
  Texas 84, 106
Petrucci, Vincent 86–7
Petznick, Earl and Melinda 176
Pheasant Ridge 62
Philip II, King of Spain 12
Phillips, Doug 216
Phillips, Madonna 108
Phillips, Sue 216
Phillips, Tony 108
Phillips Vineyard 93, 108
Phoenix Magazine 187
Phoenix producers 180–1

INDEX 255

phylloxera
 Colorado 211, 217, 221
 Texas 58, 60–1, 74, 76
Picpoul Blanc 7
 Arizona 150, 159, 162, 176
 Texas 76, 82, 83, 88, 109
Pierce, Barbara 148
Pierce, Dan 148, 165
Pierce, Michael 151, 164–5, 174
Pierce's disease 3, 74, 78, 83, 136, 152
Pike's Peak Vineyards 201
Pillsbury, Sam 137, 159, 166–7, 186
Pillsbury Vineyards 176
Pillsbury Wine Company 148, 159, 162,
 166–7, 175
Pineda, Alonso Álvarez de 55
Pinney, Thomas 57, 131–2, 200
Pinot Blanc 117
Pinot Gris
 Arizona 137, 150, 167, 181
 Colorado 209, 210, 220, 221, 222, 235
 Texas 118
Pinot Meunier 35, 36, 37
Pinot Noir
 Arizona 163, 165, 166, 180
 Colorado 209, 210, 218, 220, 222
 New Mexico 28, 35, 36, 37, 39, 41
Pizzeria Bianco, Phoenix 184
Plum Creek Cellars 201
Plum Creek Winery 216
Portra, Garrett 213
Pothier, Kris 175, 176
Powell, John Wesley 144
Power, Velmay de Wet 119
Precept Wine 36
Primitivo 32, 138
Prohibition
 Arizona 135
 Colorado 198, 200
 New Mexico 18
 Texas 57, 58, 113
Pronghorn Vineyard 159
Puckle, Adam 188
Pueblo Revolt 13, 14
pueblos 13–14, 36–7
Puesta del Sol Vineyards 218

Qualia, Frank 57, 74, 112–13
Qualia, Louis 113
Qualia, Michael 113

Qualia, Thomas ("Tommy") 113

Rancho Loma Vineyards (RLV) 111–12
Rapaura Vintners 215
Reckert, Joanna 219, 236
Reddy, Akhil 108
Reddy, Subada 108
Reddy, Vijay 108, 110
Reddy Vineyards 73, 81, 108, 110
 and Becker Vineyards 82
 and Duchman Family Winery 86
 and Haak Vineyards & Winery 114
Red Fox Cellars 216
Redstone Vineyard 221
Red Tree Ranch Vineyard 169
Reed, Robert 59, 86, 105
Refosco 29, 34, 35, 89
Reilly, Dave 85–6
Reynolds, Lori 163
Rhumb Line Vineyard 169, 176
Rhyne, Bénédicte 90
Ribolla Gialla 34
Riddle, Jasper 38–9
Riesling
 Arizona 157, 159
 Colorado 168, 210
  Four Corners 209
  Front Range 224, 225
  Grand Valley 213, 216
  West Elks 209, 219, 221
 New Mexico 28, 34, 38, 41
 Texas 106, 116
Robert Clay Vineyards 91, 93, 99, 108
Roberts, Scott 108
Rocky Mountain Association of Vintners and
 Viticulturists 201
Rocky Mountains 205
Rocky Mountain Vineyards 214
Rolling View Vineyard 165, 176
Ron Yates 97
Roosevelt, Franklin D. 23
Roosevelt, Theodore 144
Rosato 119
Rosenzweig, Anne 185
Rouas, Bettina 185
Round Mountain Vineyard 93
Roussanne
 Arizona 156, 159, 167, 180
 Colorado 208, 210
 Texas 76, 77, 88, 96, 104, 105, 109

256 THE WINES OF SOUTHWEST U.S.A.

Route 66: 47
Ruby Cabernet 86, 111, 118
Ruffentine, Brian and Megan 181
Rune Wines 147, 158, 161–2, 167, 175
R.W. Webb Winery 136–7

Saeculum Cellars 164–6, 178
Sagmor Vineyards 109–10
Sagrantino 116, 168
Salt Lick BBQ, Driftwood 123–4
Salt Lick Cellars 124
Salt Lick Vineyards 86, 108
San Antonio Missions National Historical Park
    125
San Dominique Winery 138
Sand-Reckoner Vineyards 164
Sand-Reckoner Winery 148, 168–9
Sandstone Cellars 62
Sangiovese 7
    Arizona 137, 138, 150
        Sonoita 159, 163
        Verde Valley 170, 171, 172, 173, 176,
            177, 178
        Willcox 165, 167, 168, 169
    New Mexico 26, 29, 32, 34, 35, 41
    Texas 8, 59, 76, 77
        east 118, 119
        High Plains 106, 109
        Hill Country 85, 94, 95, 102, 104
San Jacinto, Battle of 125
Santa Anna, Antonio López de 125
Santa Fe 46–51
Santa Fe Trail 15
Saperavi 74
Sauvage, Kaibab 217–18
Sauvage Spectrum 217–18
Sauvignon Blanc
    Arizona 158, 163
    Colorado 221, 222
    New Mexico 27, 32
    Texas
        east 118
        High Plains 106
        Hill Country 84, 88, 90, 95, 101
Sazón, Santa Fe 49
Schuerman, Henry 135
Schwamb, Courtney 123
Scottsdale 183–7
Seaton, Nicholas and Katy Jane 107
Semillon 32, 84, 111, 118

Seyval Blanc 179, 210
Sharrock, Mitchell 81
Shed, The, Santa Fe 49
Sheehan, Sean 37
Sheehan Winery 27, 37
Sides, Andrew 94
Siegel, Jeff 7
Sierra Bonita Vineyards 176
Slate Mill Wine Collective 71, 97–8
Smith, Tony, Erin, Kathryn, and Kristen 81
Snapp, Rod and Cynthia 178
Snowy Peaks Winery 224
Soleado Vineyards 108
Sonoita 141, 145–7
    climate 143
    history 136, 137
    producers 155–63
    travel 187–90
Sonoita Vineyards 136, 137, 147, 162–3
Southern Rail, Phoenix 185
Southold Farm + Cellar 71, 99–100, 123
Southwest U.S.A. 1–3
    caliche soils 67
    foundational dates 3
    literature 237–8
    map 5
Southwest Wine Center 149, 151, 164–5, 172,
    173–5, 178
Southwest Wine Center Block 172
Southwest Wines 19
Souzão 81, 83, 222
Spain
    Arizona history 131, 133, 144
    Colorado history 197
    literature 237, 238
    New Mexico history 11–12, 13, 14–15, 16,
        46, 51
    Texas history 55–6, 124–5
Spicewood Vineyards 97, 100–1
Staltari, William 138
Stark, Ken and Denise 32
Steak Out Restaurant & Saloon, Sonoita 189
Steese, Steve 220–2
Ste. Genevieve 105
Steinbeck, John, The Grapes of Wrath 47
Stewart, Brad 107
Stone Cottage Cellars 220
Stone House Vineyard 101
Storm Cellar, The 220–2
Stultz, Jeff 225

INDEX 257

St. Vincent 211
Sultans 198
Summit Hogue 107
Sunglow Guest Ranch, Pearce 190–1
Sutcliffe, John 227–8
Sutcliffe Vineyards 225, 227–8
Swilling, J.W. ("Jack") 132
Swilling Irrigating and Canal Company 132
Symphony 167
Syrah
  Arizona 137, 150
    Sonoita 156, 158, 162, 163
    Verde Valley 170, 171, 172, 173, 177, 180
    Willcox 166, 167, 168
  Colorado 210
    Four Corners 226, 227
    Front Range 224
    Grand Valley 208, 213
    West Elks 218
  New Mexico 29, 41
    Mesilla Valley 27
    Middle Rio Grande Valley 37
    Mimbres Valley 26, 33
    Northern New Mexico 28
  Texas
    east 118
    High Plains 105, 106, 110
    Hill Country 88, 97, 101
    west 111

Tacoparty, Grand Junction 231
Taft, William Howard 233
TAJ Cellars 225
Talbott, Harry A. 233
Talbott Farms 233, 234
Tallent, Drew 82, 108
Tallent Vineyards 93, 108
Tamaya Vineyard 36–7
Tannat
  Arizona 156, 157, 161
  Texas 76, 77
    Hill Country 81, 82, 83, 84, 93, 104
Tatum Cellars 98
Tavern Hotel, The, Cottonwood 192
Tchelistcheff, André 59, 87
temperance movement
  Arizona 135
  New Mexico 18
Tempranillo 7
  Arizona 150

Sonoita 156, 159, 160, 161
Verde Valley 171, 172, 178
Willcox 165
Colorado 216, 218, 222, 224
New Mexico 26, 27, 29, 34, 37
Texas 8, 76, 77
  east 117, 118
  High Plains 105, 106, 110
  Hill Country 88, 89, 90, 93, 95, 96, 97, 101
  south 114
  west 111, 112
Teraldego 180
Terra, Santa Fe 50
Texas 1–3, 4
  challenges 77–9
  climate 65–6
  Frost on 8
  grape growers 106–8
  grapes 74–5, 76–7, 78
  history 55–63, 124–5
  map 66
  producers
    east Texas 117–19
    south Texas 114–16
    Texas High Plains 104–10
    Texas Hill Country 81–104
    west Texas 110–13
  regions 67–8, 75–6
    physiographic 68–70
    see also Texas High Plains; Texas Hill
      Country
Texas A&M University 59, 86
Texas Alcoholic Beverage Commission 58, 62
Texas Davis Mountains 75
Texas High Plains 68, 69, 72–3
  challenges 78–9
  grapes 76
  history 59, 62
  producers 104–10
  travel 126–8
  vineyards 106–8
Texas Hill Country 68, 69, 70–2, 75
  challenges 78, 79
  Germany heritage 63
  grapes 76
  history 59, 62
  producers 81–104
  travel 121–6
  vineyards 108
  Wine and Food Festival 87

258 THE WINES OF SOUTHWEST U.S.A.

Texas Liquor Control Act 58
Texas Tech University 59, 86
Texas Wine and Grape Growers Association 62
Texoma 75–6
Teysha Cellars 104
Thomas Volney Munson Memorial Vineyard
   61
Tillie's, Dripping Springs 124
Time Market, Tucson 189
Timmons, Andy 93–4, 107
Timmons, Lauren 107
Timmons Estate Vineyard 94
Tinto Cão 92–3
Tio Pancho/Bear Vineyards 93, 108
Tondre, Joseph 15
Touriga Nacional 76, 77
   Texas Hill Country 92, 93, 95, 102
   Texas High Plains 110
Traminette 179, 224
Treaty of Hidalgo (1848) 132
Trebbiano 77, 85
Trilogy Cellars 104
Troup, Bobby 47
Tularosa Vineyards Winery 16, 17
Turley, Richard and Padte 214
Turnbull, Corey 170, 179
Turner, Chris 172
Turquoise Trail 51
T.V. Munson Viticulture and Enology Center,
      Grayson College 61
Two Rivers Winery 168

UNESCO
   Creative Cities 46
   World Heritage Sites 125
University of Arizona 135
   College of Agriculture and Life Sciences 175
University of California, Davis (UC Davis) 173
University of Texas 86
Urbanek, Seth 102
U.S. Army 56
U.S. Department of Agriculture
   Arizona Vineyard Survey 150
   Exhibition of American Grapes 61
   Texas grape varieties 76, 78
U.S. Department of Commerce, 1909 Census
      198
U.S. Environmental Protection Agency 4

Val Verde Winery 57, 74, 75, 112–13

Vargas, Don Diego de 14
Vaudeville, Fredericksburg 124
Verde Valley 141–2, 148–9
   climate 143
   history 138–9
   producers 169–80
   travel 191–3
Verde Valley Wine Consortium 163
Vermentino 7
   Arizona 150, 161, 169, 176, 180
   Texas 76, 77, 85, 86, 119
Viala, M. Pierre 60
Villard Blanc 117
Viña Sonoita 136
Vines Resort and Spa, The 157
Vineyards, The, Cornville 192
Viognier
   Arizona 137, 150, 151
      Sonoita 156, 157
      Verde Valley 170
      Willcox 166, 167
   Colorado 208, 210, 224, 227
   New Mexico 27, 29
   Texas 76, 77, 78
      east 118
      High Plains 104, 106, 109
      Hill Country 82, 88, 89, 90, 96, 97
      west 111
Vitis aestivalis 74
Vitis arizonica 152
Vitis berlandieri 60, 61, 74
Vitis cinerea 60, 74
Vitis cordifolia 60
Vitis riparia 61
Vitis rupestris 61
Vitis vinifera 1
   Black Spanish 74
   Colorado 210
   New Mexico 4, 16, 33, 56
   phylloxera 76, 217
   Pierce's disease 74
   Texas 56, 57, 59, 76
Vivác Winery 28, 31, 40–1
Vranac 150, 159

Wade, Samuel 232
Walden Retreats, Johnson City 122
Waters Winery 161
Webb, Robert 136–7, 148, 169
Webster, Todd 111

INDEX 259

Wedding Oak Winery 101–2
Weiss, Aaron 178
West Elks 208–9
  challenges 211
  grapes 210
  producers 218–22
  travel 235–7
Wharton, Tyzok 223
White, Tim 170, 172
White Sands National Park 23
Wickham, David and Teresita 17
Wildman, Frederick S. 86
Willcox 141, 147–8
  climate 143
  history 137–8
  producers 164–9
  travel 190–1
William Chris Vineyards (WCV) 71, 85, 102–4
Williamson, Robert and Laurie 111–12
Willow City Loop 126
Wilmeth, Jet 107
Wilmeth Vineyards 82
Wilms, Norman 58
Wilson, Rae 98
Wilson, Woodrow 145
Wine Country Inn, Palisade 230
Wine King 117

Winery at Holy Cross Abbey, The 224–5
Wines and Spirits 87
Winters, R.L. 117–18
Wood, Valerie and Daniel 175

Xarel-lo 150
Ximénez, Friar Lázaro 12

Yager, Joe 233
Yates, Glena 100
Yates, Ron 97, 100–1
Yavapai Community College 138
  Southwest Wine Center 149, 151, 164–5, 172, 173–5, 178
  Southwest Wine Center Block 172
Yes We Can 103
Young, Brenda 83
Young, Robert W. 83–4
Young, Rod 170, 179

Zanutta, Alessandro 172
Zinfandel
  Arizona 150, 155, 169, 178
  Colorado 198
  New Mexico 27
  Texas 100, 118
Zúñiga, Fray Garcia de 12